VOLUME ONE HUNDRED AND NINETY TWO

ADVANCES IN IMAGING AND ELECTRON PHYSICS

Analytical, Approximate-Analytical and Numerical Methods in the Design of Energy Analyzers

VOLUME ONE HUNDRED AND NINETY TWO

ADVANCES IN IMAGING AND ELECTRON PHYSICS

Analytical, Approximate-Analytical and Numerical Methods in the Design of Energy Analyzers

VICTOR S. GUROV
Ryazan State Radio Engineering University, Russia

ARMAN O. SAULEBEKOV
Kazakhstan Branch of Lomonosov Moscow State University, Astana, Republic of Kazakhstan

ANDREY A. TRUBITSYN
Ryazan State Radio Engineering University, Russia

Translated by **Mikhail A. Monastyrskiy**
Prokhorov General Physics Institute, Russian Academy of Sciences, Moscow

ELSEVIER

AMSTERDAM • BOSTON • HEIDELBERG • LONDON
NEW YORK • OXFORD • PARIS • SAN DIEGO
SAN FRANCISCO • SINGAPORE • SYDNEY • TOKYO
Academic Press is an imprint of Elsevier

Cover photo credit:
Victor S. Gurov, Arman O. Saulebekov, Andrey A. Trubitsyn
Analytical, Approximate-Analytical and Numerical Methods in the Design of Energy Analyzers
Advances in Imaging and Electron Physics (2015) 192, pp. 1–14

Academic Press is an imprint of Elsevier
225 Wyman Street, Waltham, MA 02451, USA
525 B Street, Suite 1800, San Diego, CA 92101-4495, USA
The Boulevard, Langford Lane, Kidlington, Oxford OX5 1GB, UK
125 London Wall, London, EC2Y 5AS, UK

First edition 2015

ISBN: 978-0-12-802252-8
ISSN: 1076-5670

For information on all Academic Press publications
visit our website at http://store.elsevier.com/

CONTENTS

FOREWORD

The main application of an energy analyzer of charged particles, the design of which constitutes the content of this book, is electron spectroscopy. The great variety of the tasks and methods of electron spectroscopy is the reason for the multiple requirements that are generally imposed on energy analyzer design. Nevertheless, two principal parameters can be distinguished to characterize an energy analyzer as an electron spectrometry device: namely, the order of angular focusing and the energy dispersion. The first of these parameters determines the sensitivity of the energy analyzer, while the second determines its energy resolution. The nature of these parameters is self-contradictory, which manifests itself in the fact that the attempts of improving one of them commonly lead to deterioration of the other. Thus, the key to solving the problem of optimal design of energy analyzers lies in the ways to reduce the extent of this conflict. In particular, a partial resolution of this physical contradiction may be achieved by using the electrostatic fields with low spherical aberration, or, alternatively, by creating the conditions for high-order angular and spatial focusing.

In this book, we discuss three different approaches to the analysis of the focusing and dispersive electrostatic fields that are used in energy analyzers. These comprise the analytical methods, which allow researchers to find a solution in terms of elementary mathematical functions in some cases; the approximate-analytical methods, when a solution is sought in the form of infinite series; and, finally, the numerical methods that employ all the power of the modern methods of computational mathematics and provide a solution in the form of a function defined on a mesh with a restricted number of nodes. In this context, the authors would like to state their confidence that success in the design of new types of energy analyzers undoubtedly requires a reasonable combination of all three of the approaches.

The original studies outlined in this book summarize the results of more than 20 years of work of each of the authors in the field of energy analysis, friendly discussions on all areas of electron optics, and independent examination of the devices designed. The book can be useful for students of charged particle optics, scientists and engineers involved in the development of the new advanced energy analyzers, and users of electron spectroscopy devices who want to keep abreast of current trends in this area.

All the chapters are the result of collective work; however, we should distinguish a basic contribution of each of the authors in writing the relevant chapters. Chapter 1 has a general introductory character. Chapter 2, which presents analytical methods for the design of certain types of energy analyzers, has been basically written by Professor V. S. Gurov. The main contribution to Chapter 3, dedicated to the approximate-analytical methods, was made by Professor A. O. Saulebekov. Finally, the content of Chapter 4, which outlines the numerical methods as applied to designing the energy analyzers, is primarily the work of Professor A. A. Trubitsyn.

In conclusion, the authors express their sincere respect for and deep gratitude to their colleagues and teachers who, given the level of their excellent work and general contribution to science, can be rightly regarded as the "aces" of electron and ion optics–namely, V. V. Zashkvara, S. Ya. Yavor, E. P. Sheretov, K. Sh. Chokin, B. U. Ashimbayeva, and V. A. Gorelik. Without their kind attention and fruitful contacts with the authors in the past and present, this monograph would not have been produced. Special thanks are extended to Dr. M. A. Monastyrskiy, who translated this text into English and made a number of exceedingly useful professional comments on the form and content of virtually all its sections.

We also highly appreciate the financial support of the Russian Science Foundation (grant No. 15-19-00132) in the creation of this publication.

Victor S. Gurov
Arman O. Saulebekov
Andrey A. Trubitsyn

PREFACE

The theme of the present volume is the design of energy analysers for charged particles and in particular, the various approaches that have been developed for the task: analytical methods, approximate analytical methods for more realistic field models and numerical methods.

The book opens with a survey of the traditional types of energy analyser and of more advanced models. This sets the scene for the three families of methods employed in the design of these instruments. Chapter 2 covers the analytical methods based on cylindrical and hyperbolic field models. This is followed by a chapter on more realistic field models for which the mathematical analysis requires certain approximations. Several hexapole–cylindrical analysers are studied in this way.

Useful though model fields are to establish the general behaviour of an analyser and its dependence on the various parameters present, an exact simulation requires numerical methods and these are the subject of Chapter 4. Such methods are described in considerable detail and the chapter ends with several examples of simulations of different types of analysers.

I am very grateful to the authors for preparing this clear and full account of the subject. The original Russian text has been translated by Dr Mikhail Monastyrskiy, whose name is already familiar to readers of these Advances and who has published many papers on this and related subjects. I am sure that readers will join me in thanking him for undertaking this challenging task.

Peter W. Hawkes

FUTURE CONTRIBUTIONS

S. Ando
Gradient operators and edge and corner detection

J. Angulo
Mathematical morphology for complex and quaternion-valued images

D. Batchelor
Soft x-ray microscopy

E. BayroCorrochano
Quaternion wavelet transforms

C. Beeli
Structure and microscopy of quasicrystals

C. Bobisch and R. Möller
Ballistic electron microscopy

F. Bociort
Saddle-point methods in lens design

K. Bredies
Diffusion tensor imaging

A. Broers
A retrospective

N. Chandra and R. Ghosh
Quantum entanglement in electron optics

A. Cornejo Rodriguez and F. Granados Agustin
Ronchigram quantification

L.D. Duffy and A. Dragt (Vol. 193)
Eigen-emittance

J. Elorza
Fuzzy operators

R.G. Forbes
Liquid metal ion sources

P.L. Gai and E.D. Boyes
Aberration-corrected environmental microscopy

M. Haschke
Micro-XRF excitation in the scanning electron microscope

R. Herring and B. McMorran
Electron vortex beams

M.S. Isaacson
Early STEM development

K. Ishizuka
Contrast transfer and crystal images

K. Jensen, D. Shiffler and J. Luginsland
Physics of field emission cold cathodes

M. Jourlin (vol. 194)
Logarithmic image processing, the LIP model. Theory and applications

U. Kaiser
The sub-Ångström low-voltage electron microscope project (SALVE)

C.T. Koch
In-line electron holography

O.L. Krivanek
Aberration-corrected STEM

M. Kroupa
The Timepix detector and its applications

B. Lencová
Modern developments in electron optical calculations

H. Lichte
Developments in electron holography

M. Matsuya
Calculation of aberration coefficients using Lie algebra

J.A. Monsoriu
Fractal zone plates

L. Muray
Miniature electron optics and applications

M.A. O'Keefe
Electron image simulation

V. Ortalan
Ultrafast electron microscopy

D. Paganin, T. Gureyev and K. Pavlov
Intensity-linearmethods in inverse imaging

N. Papamarkos and A. Kesidis
The inverse Hough transform

Q. Ramasse and R. Brydson
The SuperSTEM laboratory

B. Rieger and A.J. Koster
Image formation in cryo-electron microscopy

P. Rocca and M. Donelli
Imaging of dielectric objects

J. Rodenburg
Lensless imaging

J. Rouse, H.-n. Liu and E. Munro
The role of differential algebra in electron optics

J. Sánchez
Fisher vector encoding for the classification of natural images

P. Santi
Light sheet fluorescence microscopy

R. Shimizu, T. Ikuta and Y. Takai
Defocus image modulation processing in real time

T. Soma
Focus-deflection systems and their applications

I.F. Spivak-Lavrov (Vol. 193)
Analytical methods of calculation and simulation of new schemes of static and time-of-flight mass spectrometers

J. Valdés
Recent developments concerning the Système International (SI)

CHAPTER ONE

Energy Analysis of Charged Particle Flows

Contents

Energy analyzers serve to detect charged particles with prescribed characteristic energies, or, in other words, to isolate charged particles, the energies of which fall in a narrow energy interval ΔE.

1. BASIC PARAMETERS

The main consumer parameter of energy analyzers is energy resolution. Two definitions of energy resolution are used in spectroscopy (see, for example, Seah & Briggs, 1992). The first definition is associated with the absolute energy resolution ΔE, representing the entire energy range within which the particles reach the collector. Such energy resolution is called the *basic resolution*. In practice, the resolution ΔE_{FWHM}, which is defined as a full width at half maximum (FWHM) of the energy transmittance function, is often used. The second definition considers the relative energy resolution $R = \Delta E/E_0$, or $R_{FWHM} = \Delta E_{FWHM}/E_0$, where E_0 is the kinetic energy that corresponds to the average energy to which the energy analyzer is tuned. Sometimes the energy resolution is defined as a value equal to the inverse of R.

Another important characteristic of a device intended for energy analysis, which in essence determines the second consumer parameter of a spectrometer (namely, the signal-to-noise ratio), is its capacity for registering the particles emitted by different parts of an extended source. This capacity, called

Advances in Imaging and Electron Physics, Volume 192
ISSN 1076-5670
http://dx.doi.org/10.1016/bs.aiep.2015.08.001

brightness (Kozlov, 1978), can be calculated as the integral of luminosity taken over the source surface.

The luminosity of a corpuscular-optical device is equivalent to the geometrical solid angle Ω of particle collection. The so-called relative angle G is commonly introduced (Kozlov, 1978), which is defined as a ratio of the angle Ω to the full solid angle 4π: $G=(\Omega/4\pi)100\%$. Physically, the luminosity expresses a ratio of the number of particles reaching the collector to the total number of particles emitted during the same time by a monoenergetic isotropic point source located at the analyzer focus, provided that the latter is properly tuned.

Mathematically, the luminosity can be expressed through the planar tilting angles α_1 and α_2 of the generatrices of the cones $\Omega=2\pi(\cos\alpha_1-\cos\alpha_2)$, $\alpha_1<\alpha_2$.

2. MAIN TYPES OF ENERGY ANALYZERS

Two types of energy analyzers—retarding and dispersive—are used in the spectroscopy of charged particles.

One basic energy analyzer of the retarding type is a so-called quasi-spherical capacitor with grids (Figure 1; also see Woodruff & Delchar, 1986).

The operation principle of the quasi-spherical capacitor is as follows. Under the action of exciting radiation (labeled 1 in the figure; for example, of electrons emitted by the electron gun EG with energy E_{max}) the sample (2) under study, which is placed at the geometrical center of the capacitor, starts to emit the charged particles that travel along the radial trajectories (3). The grounded grid (4) creates a field-free drift space between the sample and the area of retarding. The grid (5) is supplied with a retarding voltage $-V_T$. The collector (6) is kept at the potential of 200–300 V relative

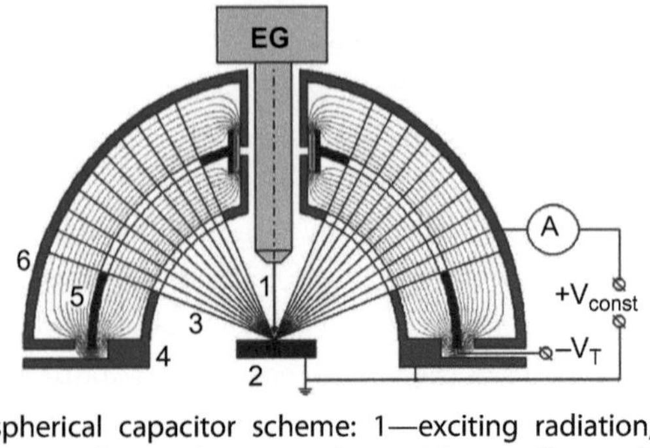

Figure 1 Quasi-spherical capacitor scheme: 1—exciting radiation, 2—sample, 3—secondary particles, 4—first grid, 5—retarding grid, 6—collector, EG—electron gun. (See the color plate.)

to the ground. The particles, with energy $E > eV_T$, reach the collector surface, so the collector current becomes equal to $I_k(E) = eN(E > eV_T)$. Here, e is the charge of the electron and $N(E > eV_T)$ is the number of particles emitted per second by the source in the energy range $[eV_T, E_{max}]$. By varying the retarding voltage from zero to a certain maximum value $V_T \leq E_{max}/e$, the dependence $I_k = I_k(V_T)$, which is called the *delay curve*, is registered. However, the delay curve has little information concerning the energy distribution of particles. The delay curve's derivative is of much bigger practical use and possesses greater information capacity.

Let $dI_k(E)$ be the elementary, infinitesimally small current created by the particles with energy E. Then $I_k(E) = \int_{E=eV_T}^{E_{max}} dI_k = e\int_{E=eV_T}^{E_{max}} dN = e\int_{E=eV_T}^{E_{max}} \frac{dN}{dE} dE$, where dN is the number of particles with energy E emitted per second. Using the rule of differentiation of integrals with a variable lower limit, we obtain

$$\frac{dI_k(E)}{dE} = -e\frac{dN(E)}{dE}. \tag{1}$$

Thus, differentiating the delay curve allows for determination of the distribution function dN/dE, and, therefore, the number of particles attributable to a narrow interval of energies ΔE: ΔN is $(dN/dE)\Delta E$.

The advantage of the analyzer of the retarding type is its high luminosity. The disadvantages include a significant level of shot noise caused by the emission of secondary electrons from the system's electrodes. The energy resolution of the capacitor with a single retarding grid reaches 2.4%, with two grids being 0.5% .

Currently, corpuscular spectroscopy most commonly employs electrostatic analyzers with energy dispersion, which in the aggregate of their analytical characteristics are far superior to retarding analyzers. The operation principle of the dispersive analyzers is based on the spatial separation of the flow of charged particles traveling in a constant electric field and the focusing of the particles with a certain energy into a point. An aperture located at the focal point transmits to the collector [a secondary electron multiplier (SEM)] the particles with a certain ratio of energy E_0 to the voltages of the analyzer electrodes in a certain energy range ΔE (the absolute resolution of the analyzer). That ratio is determined by the dispersive and focusing properties of the analyzing field, the size of the input and output apertures, analyzer geometry, analyzer manufacture quality, and some other factors. In particular, the energy measurement accuracy is influenced by external magnetic fields (including the Earth's magnetic field), because even in weak magnetic fields, the light particles are strongly deflected. Special screens made of ferromagnetic alloys (for example, μ-metal) are used to protect against stray magnetic fields.

In electron optics, a particle trajectory is determined by the ratio $E_i/V=C_i$, not by the absolute energy value E_i. Here, V is the potential of the focusing or deflection electrode of an electron-optical system (C_i=const). Thus, for each potential V, only the particles from the energy range $\Delta E=\Delta E(V)=E_2(V)-E_1(V)$, within which the value $R=\Delta E/E_0=(E_2-E_1)/E_0=(VC_2-VC_1)/VC_0=(C_2-C_1)/C_0$ remains constant, will be registered by the collector. Here, E_1 and E_2 are the minimal and maximal energies of the particles having passed through the analyzer's apertures and registered by the collector, and E_0 is the energy corresponding to the transmittance maximum. This indicates that the dispersion analyzers operate in the mode of a constant relative resolution R.

The requirement of gaining high resolution at high luminosity is contradictory. The better the focusing properties of a device and the higher the dispersion value, the less contradiction there is. Mathematically, the focusing properties can be expressed in terms of the focusing order N, which is equal to the number of first nonzero terms in the Taylor expansion of the time of arrival with respect to the initial angle α (Woodruff & Delchar, 1986):

$$x(\alpha, E_0) = x_0 + A_1(\alpha-\alpha_0) + A_2(\alpha-\alpha_0)^2 + \ldots + A_N(\alpha-\alpha_0)^N + \ldots, \quad (2)$$

where $x_0=x(\alpha_0, E_0)$ is the focus point's coordinates and α_0 is the central angle of focusing. The remaining nonzero terms define the value of spherical aberration. In particular, in the case of ideal angular focusing $A_1=A_2=\ldots A_N=0$, $N=\infty$, we get the equality $x(\alpha,E_0)=x_0$, which indicates that the time of arrival does not depend on the initial angle α.

The value D defined as the displacement of particles in case their energy is doubled represents the linear dispersion with respect to energy E (Afanasyey & Yavor, 1978):

$$D = (\partial x/\partial E)\, E. \quad (3)$$

The parameter $D^*=D/L$, where L is the characteristic size of a device, is called the *reduced (relative) dispersion*.

The Helmholtz-Lagrange theorem is of great importance in corpuscular optics employing the retarding fields (Woodruff & Delchar, 1986). If the bunch's energy before retarding is equal to γE_0, where γ is a constant, and becomes equal to E_0 after retarding, the change in the bunch's angular divergence resulted from retarding obeys the Helmholtz-Lagrange equation:

$$\Delta\alpha_0{}^2(\gamma E_0) = \Delta\alpha_1{}^2 E_0 M^2, \tag{4}$$

where $\Delta\alpha_0$ is angular divergence at energy γE_0, $\Delta\alpha_1$ is the angular divergence after retarding, and M is the linear magnification of the system.

In turn, the dispersion analyzers are subdivided into mirror-type analyzers, where the input and output of particles occur at a certain angle to the equipotential lines of the field, and deflectors, where the particles are moving virtually along the equipotential lines.

For dispersion analyzers, in contrast to retarding analyzers, the collector current at each moment of registering represents a sum of the elementary currents $dI(E)$ produced by electrons passing through the aperture with different energies in the interval $E_{min} < E_0 < E_{max}$ (Figure 2); i.e., $I_k(E_0) = \int_{E_{min}}^{E_{max}} dI(E)$.

Since $dI(E) = e \cdot dN(E)$, where $dN(E)$ is the number of electrons with the energy E passing through the aperture per second, then $I_k(E_0) = e\int_{E_{min}}^{E_{max}} dN(E)$.

By applying the well-known mean-value theorem from integral calculus to the expression $I_k(E_0) = e\int_{E_{min}}^{E_{max}} \frac{dN(E)}{dE} dE$, we obtain

$$I_k(E_0) = e\left(\frac{dN}{dE}\right)_{AV} \Delta E, \tag{5}$$

where $(dN/dE)_{AV}$ is the mean value of the function dN/dE in each act of registration, and $\Delta E = E_{max} - E_{min}$ is the analyzer's transmittance band.

Thus, based on the relationship $I_k(E_0) \sim dN/dE$, we may conclude that the collector current is proportional to the distribution function of electrons in energies.

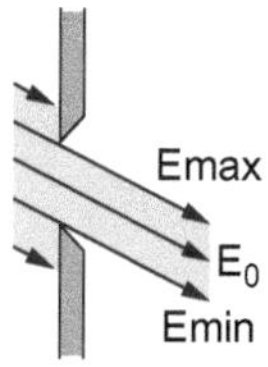

Figure 2 The passage of electrons through the output aperture.

To solve the problems of local surface analysis, when the emission source of secondary particles is small (10 μ and less), the analyzers of the cylindrical mirror type providing second-order focusing [$A_1 = A_2 = 0$ in Eq. (2)] relative to the angle of incidence (Zashkvara, Korsunskii, & Kosmachev, 1966) are frequently used.

The main elements of the analyzer (Figure 3) represent two coaxial metal cylinders 1 and 2 with different radii R_1 and R_2. The inner cylinder (labeled 1 in the figure) is usually grounded, while the external cylinder is supplied with a negative (relative to ground) potential V that can be varied over a fairly wide range. For the fringe field correction, corrective rings are mounted at the ends of cylinders, with the voltages proportional to the potential V of the outer electrode. For example, in the case of three pairs of corrective rings, their potentials should be 0.25 V, 05 V, and 0.75 V, respectively. Thus, the analyzing field is formed between the cylinders.

Secondary electrons (3) excited from the sample (4) by the primary radiation (5), such as electrons, fall into the analyzing field through a special input window (6) in the inner cylinder, covered by a fine-structure grid, and, in their motion, are deflected to the cylinder axis. The electrons that pass through the grid of the output window (7) in the inner cylinder and the restricting aperture (8) hit the collector (9), and are being registered. In the case of axis-to-axis focusing, the ratio of the analyzer tuning energy E_0 to the potential V is $\frac{E_0}{V} = \frac{1.3098}{\ln(R_2/R_1)} = \text{const}$, which indicates the opportunity to analyze the whole spectrum of energy E_0 of secondary particles by varying the potential V of the outer cylinder.

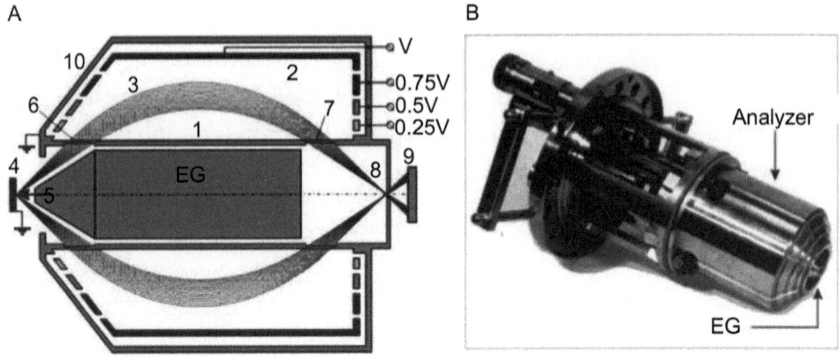

Figure 3 Electron-optical scheme (A) and external view (NITI, Ryazan) (B) of the cylindrical mirror energy analyzer: 1—inner cylinder, 2—outer cylinder, 3—secondary electrons, 4—sample, 5—primary electrons, 6—input window 7—output window, 8—diaphragm 9—collector (SEM), 10—magnetic shield, EG—electron gun. (See the color plate.)

In the case of ideal (infinitely extended cylinders) distribution of electrostatic potential between the cylinders $U(r) = V\frac{\ln(r/R_1)}{\ln(R_2/R_1)}$, $R_1 < r < R_2$, the central angle of particle input into the analyzer is $\alpha_0 = 42.3068°$, and the distance between the source and the point of focus is $6.1298R_1$. The analyzer ensures the relative energy resolution $R(\%) = \Delta E/E_0 = 1\%$ at the luminosity G about 10%.

In studying the surface properties of a solid by probing the surface with wide bunches of exciting radiation (e.g., of the X-ray type), the use of classical cylindrical mirrors is generally not advisable due to deterioration of the energy resolution R, which is caused by an opportunity for electrons, which are emitted not from the axis and possess considerably different energies from the tuning energy E_0, to pass through the system of apertures. This problem can be solved by placing a few (two, in the simplest case) successive stages of analysis and diaphragming the flow of the intermediate focuses. In this approach, the limiting aperture from the previous stage serves as a source of particles on each subsequent stage.

Figure 4 demonstrates an opportunity of using the two-stage cylindrical mirror analyzer as a research tool in studying the energy dependencies of high-emittance charged particle flows. In the scheme of this example, the particles are emitted by a large-radius source in the energy range exceeding the bandwidth of a single-stage energy analyzer.

The analyzer parameters (resolution and luminosity) in this case are virtually identical to the parameters of a single-stage mirror with a small source.

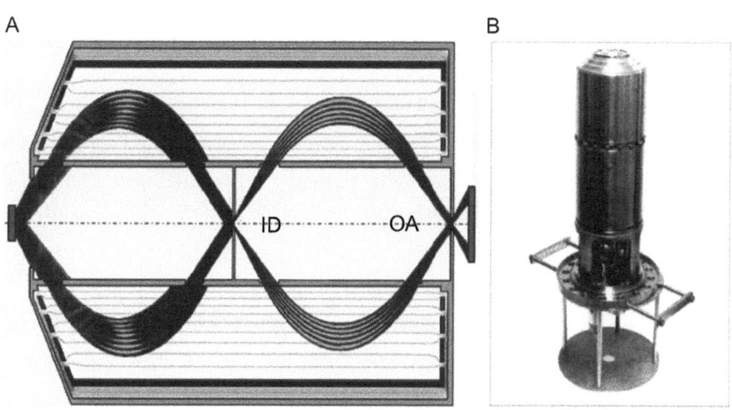

Figure 4 Electron-optical scheme (A) and external view (NITI, Ryazan) (B) of the two-stage cylindrical mirror energy analyzer: ID—intermediate diaphragm, OA—output aperture. (See the color plate.)

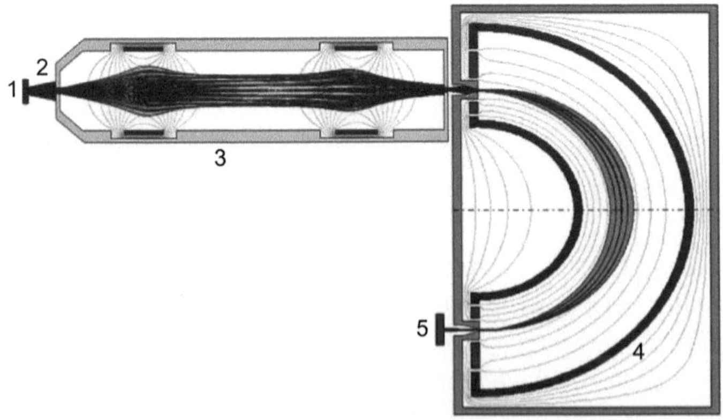

Figure 5 A hemispherical deflector (4) with the input lens system (3): 1—sample, 2—secondary (photo) electrons, 5—collector (SEM). (See the color plate.)

Much more often, in the spectroscopy problems that employ the excitation radiation in the form of bunches of large diameters, a semispherical deflector is applied, with the input lens ensuring efficient collection of secondary particles from the sample and their transportation to the analyzer's entrance (Figure 5).

The device's luminosity is approximately 100 times less than that of the cylindrical mirror, while the device emittance is 10 times higher. This is owing to high efficiency of the particle collection by the lens system; herewith, the energy resolution almost matches that of the cylindrical mirror analyzer.

3. ADVANCED ENERGY ANALYZERS

A good deal of research has been dedicated to development of theory and design in energy analysis. An energy analyzer with third-order focusing can be designed by combining two regions with different cylindrically-symmetric electrostatic fields (Franzen & Taaffe, 1983). The required field configuration is achieved by adding a third cylindrical electrode that is to be placed between the internal and outer cylinders.

Third-order focusing takes place at the input angle $\alpha_0 = 39.98°$, optimal ratio of the radii of internal cylinders equal to 1.4754, and the ratio of the field strengths on internal and outer surfaces of the intermediate cylinder equal to 1.0889. The resulting focal length along the analyzer's axis is $5.7764 \cdot R_1$, wherein R_1 is the inner radius of the cylindrical electrode.

The resolution $\Delta E/E_0$ of this analyzer reaches 0.3% within the range of input angles $\Delta\alpha = 9.34°$.

The energy analyzer proposed in Menshikov (1982) comprises an additional cylindrically shaped electrode located in the inner cylindrical electrode. Applying a negative potential to the additional electrode redirects the flow of electrons deflected by the action of electrostatic field toward the analyzer axis into the space between the inner and outer cylinders, where, in turn, the flow becomes directed toward the output aperture. This principle of analyzer design can significantly improve the device's focusing and dispersion characteristics. The relative energy resolution of the proposed device is four times higher than the classical variant of the cylindrical mirror.

The analytical method developed in Gorelik, Mashinskiy, Pikovskaya, and Protopopov (1985) shows the possibility of gaining the fourth-order focusing of the axis-ring type in a three-cascade cylindrical mirror energy analyzer of a general form. Theoretical calculations confirm more than tenfold improvement in energy resolution of this device compared to a conventional cylindrical mirror.

Some works consider the design of energy analyzers based on axisymmetric fields that are different from cylindrical ones, though also allowing analytical solutions. A unique property of spherical mirror—ideal angular focusing—had been first established in the work by Sar-El (1966). A more complete investigation of spherical mirror properties was performed by Zashkvara, Yurchak, and Bylinkin. (1988), who proposed a scheme in which energy dispersion does not depend on the input angle and equals the doubled radius of internal spherical electrode. In that scheme, the source and image are positioned at the diametrically opposite points on the internal spherical mirror surface. The device's scheme applicable for practical implementation is presented in Figure 6.

A number of papers (Ovsyannikova & Yavor, 1978; Zashkvara, Ilyin, & Kruchkov, 1977; Gurov & Sheretov, 1984; Sheretov & Gurov, 1985; Sheretov, Safonov, Gurov, 1987) are dedicated to the study of electron-optical characteristics of energy analyzers based on a hyperboloidal axisymmetric electrode system. Two modes of the device operation are considered in detail: (1) with input and output of charged particles into/out the field via the input/output end-face electrodes (Figure 7), and (2) with input and output of charged particles into or out of the field via the same end-face electrode. It is shown that implementation of the return mode in the hyperboloidal analyzer allows first- and second-order angular focusing of charged particles. In the case of second-order focusing, the analyzer provides

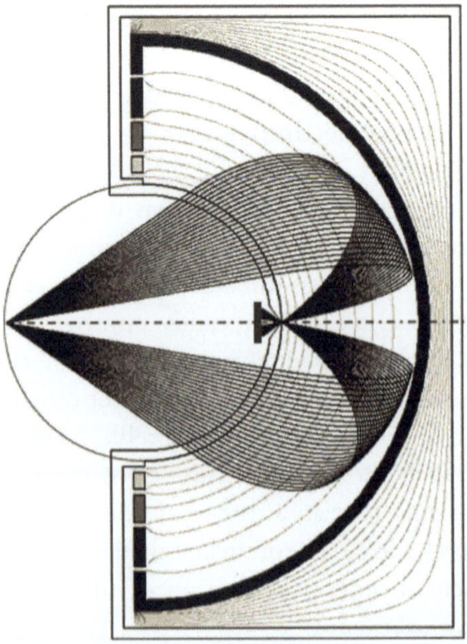

Figure 6 Electron-optical scheme of a real spherical mirror with fringe-field correction. (See the color plate.)

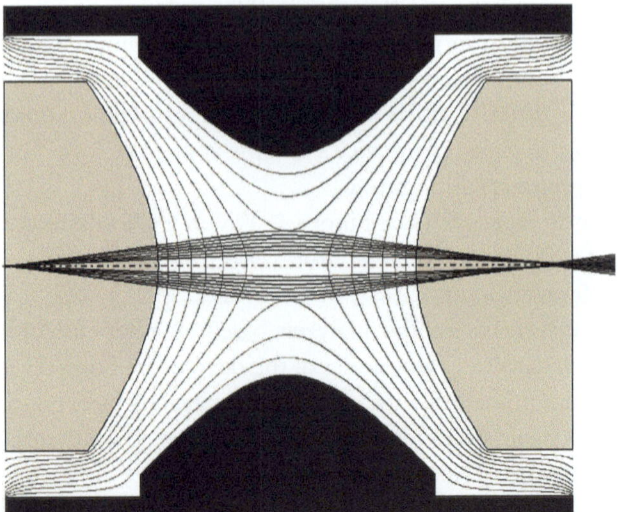

Figure 7 A scheme of the energy analyzer based on a hyperboloidal axially symmetric electrode system.

25%–30% luminosity at a resolution of 1%, which is 3–4 times higher than that of the cylindrical mirror type.

The work of Golikov, Utkin, Kholine, and Cheparuhin (1987) analyzes the dynamics of charged particles in axisymmetric Laplace fields that allow complete separation of variables in cylindrical coordinates. The most general expression for the Laplace potential with separated cylindrical coordinates r and z appears as

$$\varphi(r, z) = a \ln z + b\left(\frac{r^2}{2} - z^2\right) + cz + d, \tag{6}$$

where a, b, c, d are arbitrary real numbers.

Studies have shown that these fields may serve as a flexible basis for constructing a variety of electrode systems that provide a significant margin in dispersion, quality of focusing, and luminosity compared to the known schemes. The works by Golikov, Koltsov, and Kholine (1998, 1999) considers the possibility of designing a high-resolution, large-luminosity analyzer based on a field with quasi-conical equipotential lines (Figure 8). Numerical analysis and experimental research studies have shown that the device's resolution may vary from 0.05 to 0.12%, while the luminosity reaches a value of about 15% of 2π. The cylindrical mirror analyzer and the analyzer in question have approximately equal luminosity, while the resolution of the latter is significantly (approximately, by the order of magnitude) greater compared to the cylindrical mirror.

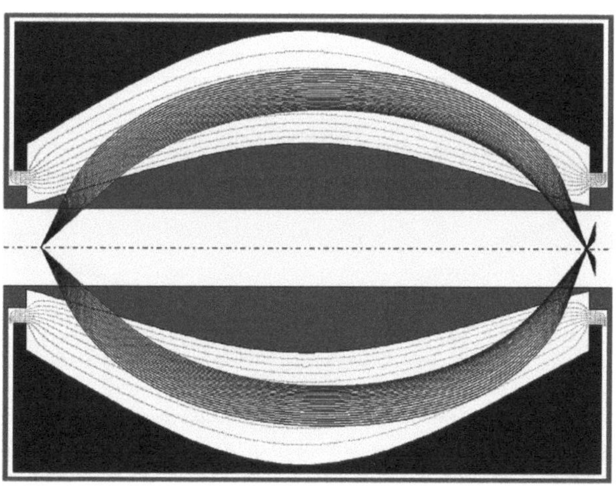

Figure 8 Scheme of a real quasi-conical energy analyzer. (See the color plate.)

A new class of axisymmetric Laplace fields proposed and studied in Ashimbaeva, Chokin, and Saulebekov (2005) and Ashimbaeva et al. (2012b) on the basis of the synthesis of multipoles and cylindrical fields is of practical interest for solving the problems of energy analysis of charged particle bunches. In the case of superposition of cylindrical mirrors and circular hexapoles, the analytical expression for potential takes the form

$$\varphi(r,z) = \mu \ln r + \gamma U_h(r,z), \tag{7}$$

where $U_h(r,z) = \frac{1}{2}\left[\ln r\left(z^2 - \frac{1}{2}r^2 - \frac{1}{2}\right) + \frac{1}{2}r^2 - \frac{1}{2}\right]$ is circular hexapoles, μ is the factor defining the weight contribution of the cylindrical field ln$r$, and γ is weight component of the circular hexapole. The analyzer scheme with end-face electrodes at $\mu = 1$ and $\gamma = -1$ (Figure 9), when the fringe field problem is absent, can be useful for practice.

Theoretically, calculated parameters of the analyzer turn out somewhat better than similar parameters of the classical cylindrical mirror.

Improving the methods of design and optimization as applied to energy analyzers of different types has been a topic of study by many authors, including Khursheed, Hoang, and Srinivasan (2012), Gorelik (2012), Fishkova (2012), Jacka, Kale, and Traitler (2003), and Cubric, Kholine, and Konishi (2011). The researchers are nowadays not virtually limited in the choice of tools for energy analysis of charged particles. However, like any other branch of human knowledge, energy analysis as a science is in

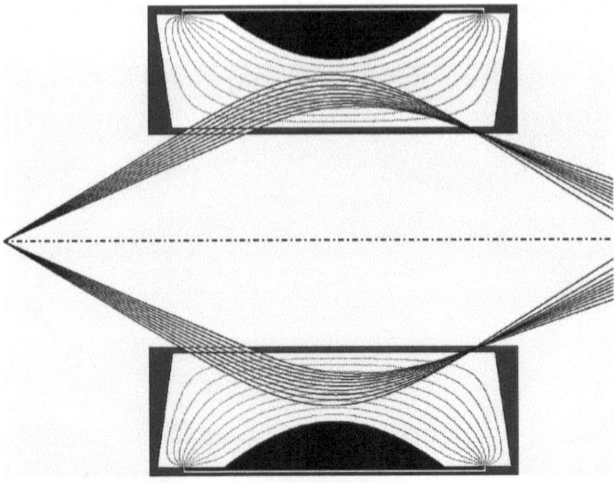

Figure 9 Physically realizable approximation of an electron-optical scheme of the hexapole cylindrical analyzer with end-face electrodes at $\mu = 1$ and $\gamma = -1$ in the axis-ring mode of focusing. (See the color plate.)

permanent development, both in terms of improving the methods of design and optimization of the tool base and from the viewpoint of developing the means for spectroscopic methods of studying the charged particle flows.

The approaches to designing the energy analyzers can be classified according to the methods of approximation of electrostatic fields and particle trajectories. Analytical methods are usable when the fields and trajectories can be expressed in terms of elementary mathematical functions. Approximate analytical methods represent a generalization of true-analytical methods for the case when the charged particle trajectories in the fields allowing analytical representation can be constructed in the form of asymptotically convergent infinite series. A set of well-developed and versatile methods of computational mathematics creates a powerful basis for numerical analysis of charged particle optics devices with virtually arbitrary geometry of electrodes. In the next three chapters, the authors demonstrate the results of studies on the development of analytical, approximate-analytical, and numerical techniques to meet the requirements of the design of energy analyzers.

CHAPTER TWO

Analytical Design Methods

Contents

Originated in the era of the formation of electron spectroscopy as a science and practice, analytical methods of designing the systems for energy analysis of charged particle flows have always played a fundamental role in the development of experimental physics and its applications.

At various times over the years, a number of devices have been designed and become widespread, including energy analyzers with retarding fields (Kozlov, 1971), energy analyzers based on Hughes-Rozhansky's cylindrical capacitor (Huges & Rojansky, 1929), on solitary electrostatic lens (Mollenstedt, 1949), as well as on electrostatic planar (Kozlov, 1978), and cylindrical (Zashkvara & Red'kin, 1969), hyperbolic (Bredov, 1959; Schmidt, 1970; and Gurov, 1976), elliptical (Shapovalov, 1968), and spherical (Purcell, 1938) mirrors. The development of these types of electrostatic energy analyzers has been stipulated by preconditioned by a wide range of problems that can be solved with their use and a multitude of requirements that are imposed on the parameters of energy analyzers as an integral part of analytical or technological equipment.

The energy analyzers with retarding fields do not employ directional focusing, which explains why they are only used to determine the longitudinal components of charged particle energies. Herewith, their energy resolution currently amounts to about 1%. Complication of the design of the energy analyzers of that type, particularly by introducing additional fine-structure grids, allows one to improve the energy resolution up to 0.5% (Afanasyev & Yavor, 1978).

Advances in Imaging and Electron Physics, Volume 192
ISSN 1076-5670
http://dx.doi.org/10.1016/bs.aiep.2015.08.002

The simplicity of design combined with sufficiently high-energy resolution contributed to expansion of the use of the Hughes-Rozhansky capacitor in experimental studies (Kozlov, 1971). The experimentally obtained energy resolution in this instrument varies from 0.3% to 2% (Huges & Rojansky, 1929).

Widespread are energy analyzers with planar electrostatic mirrors, which, due to their small size and relative design simplicity, are capable of ensuring energy resolution up to 0.3% (Kozlov, 1978). The disadvantage of such devices is their low luminosity (Afanasyev & Yavor, 1975a) that can be improved at the expense of additional focusing in the plane perpendicular to the principal trajectory by means of two complementary parallel plates (Afanasyev & Yavor, 1975b). Herewith, the normalized luminosity of that energy analyzer is quite competitive with the luminosity of the energy analyzer based on the cylindrical mirror.

However, the most popular devices are electrostatic energy analyzers based on cylindrical mirrors (Afanasyev & Yavor, 1978) because of their high-energy dispersion of the relevant electrostatic fields, their ability to implement the modes of first- and second-order spatial focusing, and axial symmetry of the field, which, taken together, can significantly improve the sensitivity of such devices with maintaining high-quality energy analysis. The experimentally obtained energy resolution in the devices of this type is about 0.2% with a luminosity of 0.01%, and about 1% with a luminosity of 8% in the modes of first- and second-order spatial focusing, respectively (Mikheeva & Protopopov, 1982).

All this has promoted the use of cylindrical energy analyzers in Auger electron spectroscopy, photoelectron spectroscopy, and electron spectroscopy for chemical analysis (Kozlov, 1978; Mikheeva & Protopopov, 1982). Gorelik, Mashinskiy, Protopopov, and Shapiro, (1976) proposes a comparison of the known types of cylindrical energy analyzers with second-order spatial focusing based on studying the relationship between the input aperture magnitude and energy resolution. It is shown that the optimally designed energy analyzer represents a device that implements the axis-axis mode of focusing with annular apertures in the minimal cross-section region. Such an energy analyzer provides an energy resolution of 0.1% at the input aperture of 0.59 sr, which corresponds to the luminosity of 4.7%. Apparently, this type of energy analyzer is one of the best in terms of its parameters among the known similar devices.

Elliptical mirrors have not gained widespread application because of the complexity of the geometry of field-generating electrodes compared to that of the cylindrical mirror analyzer, although the elliptical mirror possesses a somewhat higher specific energy dispersion (Afanasyev & Yavor, 1978).

The prospects of using planar hyperbolic electrostatic field as a basis for the design of a spectrometer of charged particles were analyzed for the first time by M. M. Bredov (1959). However, energy analyzers of that type have low luminosity, despite the attractive dispersing properties of the hyperbolic field and the opportunity for implementation of the mode of second-order directional focusing (Afanasyev & Yavor, 1978). This can be explained by the fact that the electric field being formed is not axisymmetric. The presence of the symmetry axis gives rise to a fundamentally new, three-dimensional (3D) hyperbolic electrostatic field with a quadratic potential distribution in all three coordinates.

The advantages of axisymmetric hyperbolic fields as applied to the design of efficient energy analyzers of charged particles have been outlined by several studies (Gurov, 1976; Zashkvara, Ilyin, & Kruchkov, 1977; Ovsyannikov & Yavor, 1978). One of the first significant steps in this direction was Gurov (1976), in which electron-optical and consumer parameters of the energy analyzer based on a hyperbolic axisymmetric lens were analytically derived. It is important to emphasize that considerable inhomogeneity of the potential distribution has allowed researchers to surpass the energy resolution of a cylindrical mirror at the same luminosity (Sheretov, Safonov, & Gurov, 1987).

The use of 3D inhomogeneous electrostatic fields has sparked further development of axisymmetric devices for the energy analysis of charged particles. A significant contribution to improving the quality characteristics of energy analyzers through the use of nonuniform fields and creating the unique analytical approaches and methods was provided by Russian scientists S.Ya. Yavor, S.K. Golikov and their scientific schools (see, for example, Afanasyev & Yavor (1978), Baranova & Yavor (1986), Golikov, Koltsov, & Kholine (1998), Golikov, Utkin, Kholine, & Cheparuhin (1987)).

1. FOCUSING PROPERTIES OF ELECTROSTATIC MIRRORS WITH IDEAL CYLINDRICAL FIELDS

Figure 1 shows the meridional cross section of the upper half of the cylindrical capacitor comprising two coaxial cylinders with radii R_1 and R_2. The first of these is grounded, while the other has the potential V. A point isotropic source of positively charged particles is located on the symmetry axis $0z$. The particle trajectories represent straight lines in the field-free space

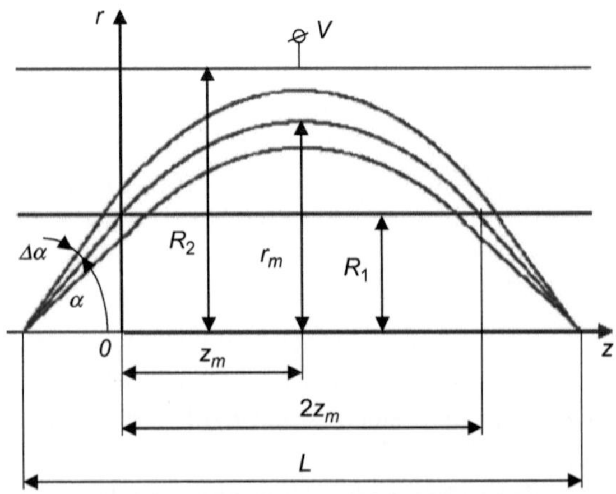

Figure 1 Particle trajectories in a cylindrical capacitor. (See the color plate.)

between the symmetry axis and inner cylinder surface. When injected into the inhomogeneous field area between the cylinders, the particles describe some curved trajectories that are symmetrical relative to the point of their maximum deviation from the symmetry axis. Since the field component along the $0z$ direction is zero, the angle of trajectory intersection with the symmetry axis is equal to the relevant angle at the starting point.

Axial symmetry of the system ensures an absolutely identical nature of the particle trajectories that start from the symmetry axis and move in the meridional plane. Consider a particle emitted from the source with initial energy E at angle α to the axis $0z$. The energy conservation law applied to the radial component of motion in the field region gives

$$\frac{m\dot{r}^2}{2} + qV\frac{\ln r/R_1}{\ln R_2/R_1} = C = const, \tag{1}$$

where q and m are the charge and mass of the particle. The motion uniformity along the symmetry axis defines the expression for the corresponding projection $z = (v_0 \cos\alpha)t$, or $dz = (v_0 \cos\alpha)dt$, where $v_0 = \sqrt{2E/m}$ is the initial velocity of the particle, which is obviously equal to the particle velocity at the point of intersection of the inner cylinder surface. By means of transition from dt to dz in Eq. (1), we obtain

$$\frac{dr}{dz} = r' = \sqrt{\frac{C}{E\cos^2\alpha} - b^2 \ln\frac{r}{R_1}}, \tag{2}$$

where $b^2 = \frac{qV}{E\cos^2\alpha \cdot \ln\frac{R_2}{R_1}}$. The constant C is defined by the initial conditions on the inner cylinder surface: $r' = \text{tg}\alpha$ at $r = R_1$. Finally, we have

$$r' = \sqrt{\text{tg}^2\alpha - b^2 \ln\frac{r}{R_1}} = b\sqrt{p^2 - \ln\frac{r}{R_1}}, \tag{3}$$

where

$$p = \sqrt{\frac{E}{qV}} \ln\frac{R_2}{R_1}\sin\alpha. \tag{4}$$

The point (r_m, z_m) of the trajectory's maximal deviation from the symmetry axis can be determined from the condition $r' = 0$, which, according to Eq. (3), gives

$$r_m = R_1 e^{p^2}. \tag{5}$$

By integrating Eq. (3), we obtain the coordinate z_m:

$$z_m = \frac{1}{b}\int_{R_1}^{r_m} \frac{dr}{\sqrt{p^2 - \ln\frac{r}{R_1}}}. \tag{6}$$

Changing the variable of integration in Eq. (6) as follows:

$$p^2 - \ln\frac{r}{R_1} = x^2. \tag{7}$$

finally gives

$$z_m = \frac{2R_1 e^{p^2}}{b}\int_0^p e^{-x^2}\,dx = \frac{2R_1}{b}\theta(p), \tag{8}$$

where

$$\theta(p) = e^{p^2}\int_0^p e^{-x^2}\,dx. \tag{9}$$

It should be noted that the latter integral in Eq. (9) can be expressed through the standard probability integral $f(x) = \frac{1}{\sqrt{2\pi}}\int_0^x e^{-\frac{t^2}{2}}dt$.

Due to the symmetry of the trajectory relative to the point (z_m, r_m) of maximal deviation from the axis $0z$, the distance L from the source, at which the trajectory intersects the axis, appears as

$$L(p)=2R_1\text{ctg}\alpha+2z_m=2R_1\text{ctg}\alpha[1\ +\ 2p\theta(p)]. \tag{10}$$

Expanding the function L into the Taylor series in powers of α, we obtain

$$\begin{aligned}L&=L_0+\frac{\partial L}{\partial\alpha}\Delta\alpha+\frac{1}{2!}\frac{\partial^2 L}{\partial\alpha^2}(\Delta\alpha)^2+\frac{1}{3!}\frac{\partial^3 L}{\partial\alpha^3}(\Delta\alpha)^3+\ldots=\\&=L_0+A_I\Delta\alpha+A_{II}(\Delta\alpha)^2+A_{III}(\Delta\alpha)^3+\end{aligned} \tag{11}$$

The condition of Nth-order focusing signifies that the first N terms in Eq. (11) vanish:

$$A_I=A_{II}=A_{III}=\ \ldots\ =A_N=0. \tag{12}$$

A chain of rather obvious relations

$$\begin{aligned}&\frac{\partial p}{\partial\alpha}=p\cdot\text{ctg}\alpha,\ \frac{d\theta}{dp}=1+2p\theta,\ \frac{\partial\theta}{\partial\alpha}=\frac{d\theta}{\partial p}\frac{\partial p}{\partial\alpha},\ \frac{\partial(p\theta)}{\partial\alpha}\\&\quad=p\cdot\text{ctg}\alpha[\theta+p(1+2p\theta)]\end{aligned} \tag{13}$$

brings us to the expressions for the first two expansion coefficients:

$$A_I=2R_1\left[2p^2(1+2p\theta)\text{ctg}^2\alpha-2p\theta-\text{ctg}^2\alpha-1\right], \tag{14}$$

$$A_{II}=2R_1\text{ctg}\alpha\left[1+\text{ctg}^2\alpha-3p^2+2p^4\text{ctg}^2\alpha+\left(2p^2\text{ctg}^2\alpha-6p^2+4p^4\text{ctg}^2\alpha-1\right)p\theta\right] \tag{15}$$

$$\begin{aligned}A_{III}=&\frac{2R_1}{3}\left[-\left(1+3\text{ctg}^2\alpha\right)\left(1+\text{ctg}^2\alpha\right)\right.\\&-\left(4\text{ctg}^2\alpha-4p^2\text{ctg}^4\alpha+12p^2\text{ctg}^2\alpha-4p^4\text{ctg}^4\alpha-3\right)p^2\\&\left.+\left(1-20p^2\text{ctg}^2\alpha+6p^2+12p^4\text{ctg}^4\alpha-24p^4\text{ctg}^2\alpha+8p^6\text{ctg}^4\alpha\right)p\theta\right]\end{aligned} \tag{16}$$

Thus, the condition $A_I=0$ of first-order focusing takes the following form:

$$\text{ctg}^2\alpha=\frac{1+2p\theta(p)}{2p^2+4p^3\theta(p)-1}. \tag{17}$$

Substituting Eq. (17) into Eq. (15), and the subsequent numerical solving of the equation $A_{II}(p)=0$ (see Figure 2a) give the value $p=p_0\approx 0.77$, which ensures the fulfillment of the second-order focusing condition $A_I=A_{II}=0$.

By using Eq. (4), Eq. (17), and the abovementioned value of $p=p_0$, we can determine the values of the angle $\alpha_0=42.3068°$ and the normalized

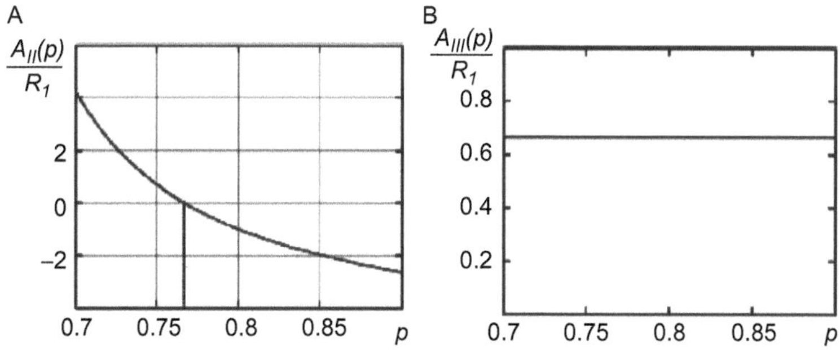

Figure 2 The second (A) and third (B) coefficients of Eq. (11) as functions of the parameter p with the first-order focusing condition $A_I = 0$ being fulfilled. (See the color plate.)

energy of particles $K_0 = \frac{E}{qV} \ln\frac{R_2}{R_1} = \left(\frac{p_0}{\sin\alpha_0}\right)^2 = 1.3098$, which ensure the mode of second-order focusing. In this mode, in accordance with Eq. (10), the distance L between the source and focus constitutes 6.1298 R_1.

The third term A_{III} in Eq. (16) equals $0.66R_1$ at $\alpha = \alpha_0$ and $p = p_0$ and turns out to be different from zero (see Figure 2 b). This represents direct evidence of the principal impossibility of third-order focusing in a cylindrical field.

The capability of focusing the charged particles determines the device luminosity (sensitivity of spectral analysis) for given restricting apertures; however, it does not reflect the capability of the device to spatially separate the particles according to their initial energies. Here, we need to evaluate the energy dispersion that determines the basic consumer-oriented parameter of the energy analyzer—its resolving power.

Using the definition of linear dispersion,

$$D = \frac{\partial L}{\partial E} E = \frac{\partial L}{\partial p}\frac{\partial p}{\partial E} E, \tag{18}$$

and performing relevant operations of differentiation in Eqs. (10) and (4) under conditions of second-order focusing [in other words, taking into consideration Eq. (17) at $p = p_0$, $\alpha = \alpha_0$], we obtain

$$D = 2R_1 \mathrm{ctg}\alpha_0 p_0 \left\{p_0 + \left[1 + 2p_0^2\theta(p_0)\right]\right\} = 5.6R_1. \tag{19}$$

Eq. (18) allows approximate estimation of the energy resolution value:

$$R = \frac{\Delta E}{E} \approx \frac{\Delta L}{D} = \frac{2R_d \mathrm{ctg}\alpha_0}{5.6R_1} \approx 0.4\frac{R_d}{R_1}, \tag{20}$$

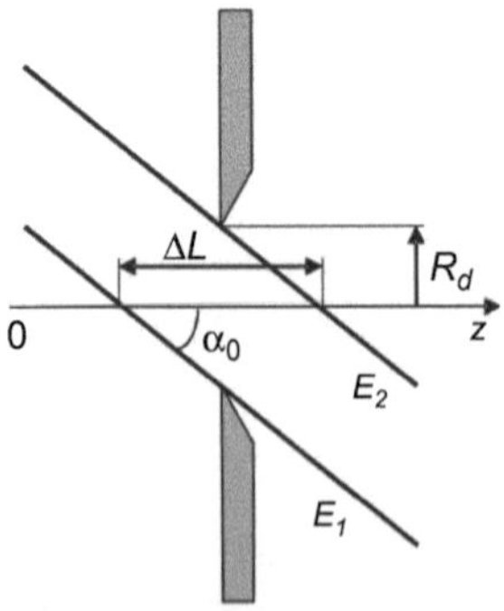

Figure 3 Estimation of the energy resolution.

where ΔL is the distance between the points of axis intersection by the trajectories with different initial energies in the interval $\Delta E = E_2 - E_1$ and initial angle $\alpha \approx 42°$ of the particle exit from the source. The distance ΔL is determined by the radius R_d of the output diaphragm (Figure 3).

Obtaining strict definitions of energy resolution and luminosity, as well as establishing the relationship between these concepts, are based on the analysis of the response function (instrumental function) of the system. One of the ways of calculating the instrumental function is based on constructing the analyzer transmittance region in the coordinates (K, α).

The transmittance region is defined as follows. First, the curve $K_1(\alpha)$ is calculated to describe the relationship between the angle α and normalized energy K of electrons, the trajectories of which touch the lower edge of the circular output diaphragm. Then, the similar curve $K_2(\alpha)$ is calculated for electrons, the trajectories of which touch the opposite edge of the diaphragm. Obviously, the electrons having the energy K in the interval $K_1(\alpha) \leq K(\alpha) \leq K_2(\alpha)$ pass through the diaphragm and are registered by the collector. The operation range of the angles $\alpha_{min} \leq \alpha \leq \alpha_{max}$ is determined by input and output windows in the inner cylindrical electrode, so the analyzer transmittance region with fixed dimensions of input and output apertures is described by the system of inequalities $\alpha_{min} \leq \alpha \leq \alpha_{max}$,, $K_1(\alpha) \leq K(\alpha) \leq K_2(\alpha)$, while its basic resolution and luminosity can be calculated as $R_B = (K_{max} - K_{min})/K_0 \times 100\%$ and $\Omega = \Omega_0/2\pi \times 100\% = (\cos\alpha_{min} - \cos\alpha_{max}) \times 100\%$, respectively. Here, K_{max} and K_{min} are the maximal and minimal energies of the transmittance region, K_0 is the analyzer's tuning energy, and Ω_0 is the solid angle of collecting the electrons having the energy K_0.

It should be noted that the optimal angle range $[\alpha_{min}, \alpha_{max}]$ is defined for each particular size of the output diaphragm—namely, the left point α_{min} and right point α_{max} are defined as the points of intersection of the straight line $K=K_0$, with the curves $K=K_2(\alpha)$ and $K=K_1(\alpha)$, respectively. The range $[\alpha_{min}, \alpha_{max}]$ is optimal because its decrease leads to a decrease in the analyzer's luminosity, with the top of the instrumental function flattened, while the range increase results in resolution deterioration without an increase in luminosity; i.e., the broadening of the instrumental function broadening occurs.

According to this definition, instrumental function represents the intensity dependence of a monochromatic flow of charged particles, emitted from a source with homogeneous luminosity and passed through the receiving slits, on the analyzer's tuning energy (Afanasyev & Yavor, 1978). The instrumental function's graph allows the determination of the angular range $(\alpha_{min}, \alpha_{max})$ belonging to the transmittance region $K_1(\alpha) \leq K(\alpha) \leq K_2(\alpha)$ for each energy value in the interval $K_{min} \leq K \leq K_{max}$; i.e., to determine the dependence $\Omega = \Omega(K)$, where $\Omega = 2\pi(\cos(\alpha_K)_{min} - \cos(\alpha_K)_{max})$.

Figure 4a shows the transmittance region, while Figure 4b displays the instrumental function of the cylindrical energy analyzer, providing (as noted previously) a mode of second-order focusing of the axis-axis type. The output aperture with radius $R_d = 0.017R_1$ is located at the focus point. The optimal range of angles is $42° \pm 6°$. The analyzer luminosity constitutes 7% from 4π, the basic energy resolution is $R_B = 1.3\%$, and the relative full width at half maximum (FWHM) value of the instrumental function is $R_{FWHM} = 0.6\%$.

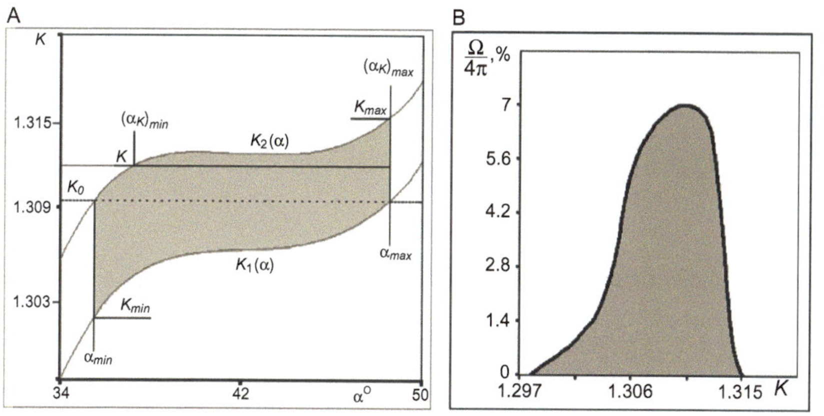

Figure 4 Transmittance region (A) and instrumental function (B) of the energy analyzer in the mode of second-order focusing. (See the color plate.)

With minor changes, the methodology of searching the focusing conditions of the axis-axis type can be applied to the general cases of axis-ring and ring-axis focusing (Zashkvara & Red'kin, 1969).

2. FOCUSING PROPERTIES OF ELECTROSTATIC MIRRORS WITH IDEAL HYPERBOLIC FIELDS

Hyperbolic mirrors employ two-dimensional (2D; Bredov, 1959) and 3D (Gurov, 1976; Zashkvara et al., 1977; Ovsyannikov & Yavor, 1978) electrostatic fields with quadratic potential distribution, which commonly possess axial symmetry. Planar hyperbolic fields are generated by a system of linear electrodes with a hyperbolic cross-sectional shape. 3D hyperbolic fields can be formed by various combinations of electrode systems representing one- or two-sheet hyperboloids of revolution. In the simplified electrode systems, one or two hyperbolic electrodes may be replaced by the electrodes, the surfaces of which coincide with relevant asymptotic surfaces.

Let us derive analytical expressions for the field-generating electrodes of the hyperboloidal electrode systems. In general, the potential distribution generated by a hyperboloidal electrode system can be represented in the following form Sheretov, Gurov, Safonov, & Philippov (1999):

$$\varphi(x, y, z) = \frac{\varphi_x}{1 + p_0 + n_0}\left\{1 + p_0 + k_0 n_0 + \frac{1 - k_0}{d_a^2}\left[x^2 - (1 + p_0)z^2 + p_0 y^2\right]\right\}, \tag{21}$$

where $n_0 = \frac{x_a^2}{d_a^2}$ and $p_0 = \frac{x_a^2}{y_a^2}$ are the geometric coefficients defining the electrode system's type; x_a, y_a, and d_a are the minimal distances from the electrode system's center to the electrodes along relevant axes; and $k_0 = \frac{\varphi_z}{\varphi_x}$; $\varphi(x_a, 0, 0) = \varphi(0, y_a, 0) = \varphi_x$; $\varphi(0, 0, d_a) = \varphi_z$; and φ_x, φ_z are the potentials of the analyzer's electrodes (Figure 5).

The equation of the electrode surface carrying the potential φ_x appears as

$$n_0 d_a^2 = x^2 - (1 + p_0)z^2 + p_0 y^2. \tag{22}$$

For the electrode with potential φ_z, we get

$$-(1 + p_0)d_a^2 = x^2 - (1 + p_0)z^2 + p_0 y^2. \tag{23}$$

By varying the coefficients n_0 and ρ_0, we can obtain the entire manifold of hyperboloidal systems generating 2D and 3D electrostatic fields with quadratic potential distributions.

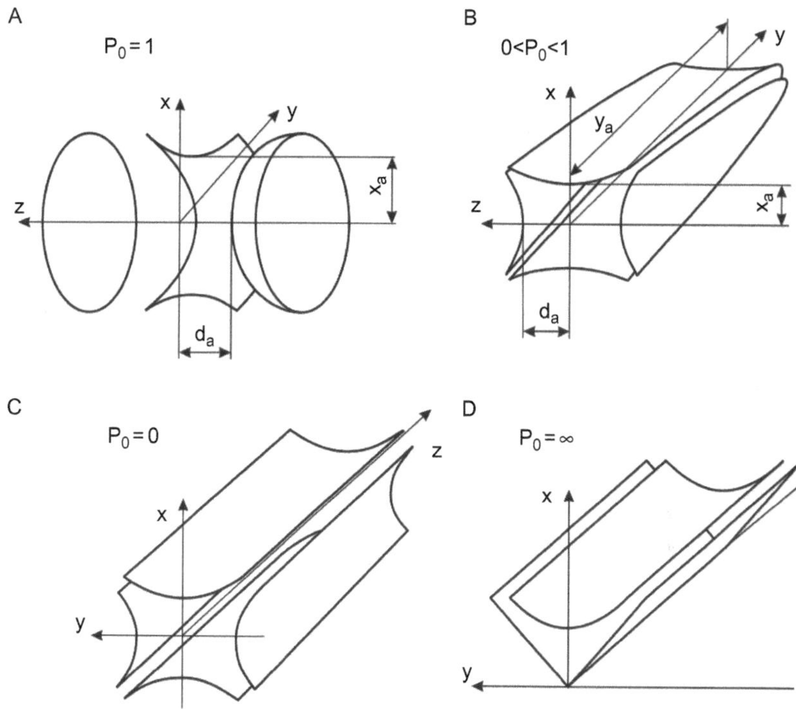

Figure 5 Main types of hyperboloidal electrode systems generating electrostatic fields with quadratic potential distributions: A—axisymmetric, B—elliptical, C—quadrupole, D-monopole.

For electrode systems generating 3D hyperbolic fields, the values x_a, y_a, and d_a have equal orders of magnitude, which means that the coefficients p_0 and n_0 are close to unit$=1$. The coefficient k_0 is commonly chosen to be equal to zero, though one can assign $k_0=-1$ to increase the field strength. If we put $p_0=1$ into Eqs. (22) and (23), this system of equations will describe the axisymmetric hyperboloidal electrode system capable of generating an axially symmetric hyperbolic field (Sheretov et al., 1999; also see Figure 5a). The field-generating electrodes of the hyperboloidal electrode systems that form 3D electrostatic fields with $0 < p_0 < 1$ have ellipses in the cross section perpendicular to the longitudinal axis $0y$ (Figure 5b).

In the case of $\rho_0=0$, the longitudinal field along the axis $0y$ is zero, and we obtain a linear quadrupole electrode system (Figure 5c), which has found widespread application in electron optics (Baranova & Yavor, 1986). Usually, the quadrupole electrode systems have $n_0=1$ and $k_0=-1$.

In the limit $\rho_0 \rightarrow \infty$, Eq. (22) describes an asymptotic corner electrode; herewith, the entire system [Eqs. (22) and (23)] represents the so-called monopole—

a fourth part of the quadrupole electrode system (von Zahn, 1963; also see Figure 5d). The corner electrode of a monopole electrode system may have a hyperbolic profile (Sheretov et al., 1999).

Quadrupole and monopole electrode systems represent the sources of planar and hyperbolic electrostatic fields.

2.1 2D Hyperbolic Electrostatic Fields

The focusing properties of 2D planar hyperbolic electrostatic fields were investigated for the first time by M. M. Bredov (1959). In the energy analyzer designed by Bredov, a planar hyperbolic field is formed by two cylindrical hyperbolic surfaces. The energy analyzer operates in mirror mode; the charged particles are being introduced into the field region and left at an acute angle through the slits made in one of the electrodes. Herewith, a retarding potential (relative to the sign of charged particles) is applied to the opposite electrode.

Let us obtain the conditions for spatial focusing of charged particles in a planar hyperbolic field. Consider the charged particle motion in a planar hyperbolic field whose potential is given by the relation

$$U(x, y, z) = a\left(x^2 - y^2\right) + bx + c. \tag{24}$$

Obviously, the potential φ meets the Laplace equation $\Delta\varphi = 0$, and relevant equipotential surfaces represent hyperbolic cylinders.

Assume that a charged particle with mass m and charge q, with the initial velocity components $\dot{x}_0$, $\dot{y}_0$, and $\dot{z}_0$ at the initial moment $t = 0$, finds itself at the point x_0, y_0, z_0 on the surface $\varphi = 0$, which separates the free-field region from the region with the potential described by Eq. (24).

Integrating the equations

$$\begin{aligned} m\ddot{x} &= -qE_x \\ m\ddot{y} &= -qE_y \\ m\ddot{z} &= -qE_z, \end{aligned} \tag{25}$$

where

$$\begin{aligned} E_x &= \frac{\partial\varphi(x, y, z)}{\partial x} = 2ax + b \\ E_y &\frac{\partial\varphi(x, y, z)}{\partial y} = -2ay \\ E_z &= \frac{\partial\varphi(x, y, z)}{\partial z} = 0, \end{aligned} \tag{26}$$

we obtain the following explicit equations describing the charged particle motion in the hyperbolic field [Eq. (24)]:

$$x = \frac{b}{2a}(\mathrm{ch}kt - 1) + x_0 \mathrm{ch}kt + \frac{\dot{x}_0}{k}\mathrm{ch}kt \tag{27}$$

$$y = y_0 \cos kt + \frac{\dot{y}_0}{k}\sin kt \tag{28}$$

$$z = z_0 + \dot{z}_0 t, \tag{29}$$

where

$$k = \sqrt{2a\frac{e}{m}}.$$

Focusing in the principal trajectory plane. Consider a particle's trajectory, whose velocity vector $\vec{v}$ at the time moment $t = 0$ belongs to the plane of field symmetry and makes the angle α with the symmetry axis $0z$, so we can put

$$\begin{aligned} x_0 &= y_0 = z_0 = 0 \\ \dot{y}_0 &= 0 \\ \dot{z}_0 &= v\cos\alpha \\ \dot{x}_0 &= v\sin\alpha. \end{aligned} \tag{30}$$

Eqs. (27) and (28) take the form

$$x = \frac{b}{2a}(\mathrm{ch}kt - 1) + x_0 \mathrm{ch}kt + \frac{v}{k}\sin\alpha\,\mathrm{ch}kt \tag{31}$$

$$y = 0 \tag{32}$$

$$z = v\cos\alpha t. \tag{33}$$

The turning point with respect to the motion along the x-axis is defined by the condition

$$\dot{x}(kt_m) = 0, \tag{34}$$

or

$$\mathrm{th}(kt_m) = -\frac{2av\sin\alpha}{b\quad k} = -\beta \tag{35}$$

In accordance with Eqs. (33) and (35), the z_m-coordinate of the turning point appears as

$$z_m = -\frac{b}{2a}\beta \cdot \mathrm{ctg}\alpha \cdot \mathrm{arcth}\beta. \tag{36}$$

Consider now a point source of the particles located in the equipotential region at the distance H from the symmetry axis (Figure 6).

Consider for now the trajectories belonging to the plane of field symmetry. Due to symmetry relative to the plane $z = z_m$ of the trajectory outgoing from the point O at the angle α to the axis Oz, the particle crosses the line $x = -H$ for the second time at a distance L from the point O equal to

$$L = 2\left(H\mathrm{ctg}\alpha - \frac{b}{2a}\mathrm{ctg}\alpha \cdot \beta \cdot \mathrm{arth}\beta\right). \tag{37}$$

Denote

$$\frac{2a}{b}H = \varepsilon, \tag{38}$$

then

$$L = \frac{b}{a}\mathrm{ctg}\alpha(\varepsilon - \beta\mathrm{arth}\beta). \tag{39}$$

Let us now satisfy the requirement of second-order focusing at the point O′ of the particles emitted from the point O at the angle α:

$$\frac{\partial L}{\partial \alpha} = 0, \; \frac{\partial^2 L}{\partial \alpha^2} = 0. \tag{40}$$

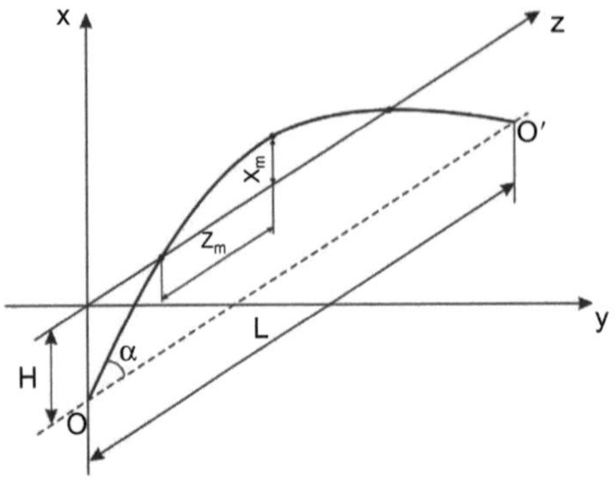

Figure 6 Characteristic parameters of the charged particle's trajectory in a planar hyperbolic field.

After differentiating Eq. (38) and performing simple transformations, we obtain

$$\varepsilon - \beta \mathrm{arth}\beta \cdot \sin^2\alpha + \frac{\beta^2}{1-\beta^2}\cos^2\alpha = 0 \tag{41}$$

$$2\frac{\beta}{\left(1-\beta^2\right)^2}\cos^2\alpha - 3\left(\beta \mathrm{arth}\beta + \frac{\beta^2}{1-\beta^2}\right)\sin^2\alpha = 0.$$

From the latter equation,

$$\mathrm{tg}^2\alpha = \frac{2}{3}\frac{\beta}{\left(1-\beta^2\right)^2\left(\mathrm{arth}\beta + \frac{\beta}{1-\beta^2}\right)}. \tag{42}$$

Substituting Eq. (42) into Eq. (41) finally gives

$$\varepsilon = \beta^2 \frac{\left(3\beta^2 - 1\right)\mathrm{arth}\beta - 3\beta}{3\left(1-\beta^2\right)^2 \mathrm{arth}\beta + 5\beta - 3\beta^3}. \tag{43}$$

Focusing in the plane perpendicular to the principal trajectory. Let us now consider a trajectory of the particle emitted from the point O at a small angle γ (the angle between the projection of the particle's velocity onto the plane $x=0$ and the axis $0z$) to the symmetry plane of the field. It is easy to show that the projections of such type of trajectories onto the symmetry plane of the field within the accuracy of first-order terms with respect to γ is described by the same relations as the trajectories belonging to the symmetry plane, if the angle α is treated as the angle between the velocity projection onto the plane of symmetry and axis $0z$. This means that, with the same degree of accuracy, the conditions in Eqs. (42) and (43) can be applied to the trajectories of this type.

Consider the motion equation [Eq. (28)] in this approximation. We have the obvious relationship between y_0 and $\dot{y}_0$:

$$\dot{y}_0 = \frac{y_0}{H} v_0 \sin\alpha. \tag{44}$$

To ensure focusing of particles along the axis $0y$, it is suffice to satisfy the equality $y_1 = y_0$, or the equivalent equality

$$\dot{y}_1 = -\dot{y}_0, \tag{45}$$

where y_1 and $\dot{y}_1$ are the particle's coordinate and velocity at the exit from the field region. From the other hand, the time moment of the particle's exit from the field region is obviously defined by the equality

$$t_1 = 2t_m, \tag{46}$$

where t_m is the same as in Eq. (34).

Substituting Eq. (45) into Eq. (28) with regard to Eqs. (46), (44), (38), and (35) gives

$$\varepsilon = -\frac{\beta}{\text{tgarth}\beta}. \tag{47}$$

Thus, the β value that ensures simultaneous fulfillment of the requirements of second-order focusing by α and first-order focusing by γ can be defined from the equation being a consequence of Eqs. (43) and (47):

$$3\left(1-\beta^2\right)^2 \text{arth}\beta + 5\beta - 3\beta^3 - \beta\text{tgarth}\beta\left[\left(1-3\beta^2\right)\text{arth}\beta + 3\beta\right] = 0$$

The root of this equation is $\beta = -0.8438$, whence $\varepsilon = -0.2954$ and $\alpha = 51^\circ 55'$.

Thus, exploring the focusing properties of hyperbolic planar field, Bredov has shown the possibility of second-order focusing in the plane of principle trajectory; i.e., in the plane of field symmetry $y=0$, and the possibility of first-order focusing in the plane perpendicular to the principal trajectory. In the case of the point source distanced from the input electrode by the value H, the optimal angle of particle input into the field constitutes $\alpha_{opt} = 51^\circ 55'$. In this case, the trajectory's maximum height x_m in the field, the optimal distance L between the source and focus, and the focal line length S_0 along the z-axis are defined as

$$x_m = 1.568H,\ L = 7.092H,\ S_0 = 1.275\alpha^3. \tag{48}$$

The energy analyzer dispersion along the input axis $0z$ is proportional to the ratio b/a (since this ratio determines the geometry of hyperboles–equipotentials, and, consequently, the degree of electric field inhomogeneity) and constitutes

$$D_\omega = 1.314L. \tag{49}$$

Optimizing the parameters of Bredov's parabolic mirror conducted by the authors of the work (Afanasyev & Yavor, 1975c) have led to the following energy analyzer characteristics: the half-opening of the angles in the principal trajectory plane $\Delta\alpha = 0.0677$; the half-opening of the angles in the plane perpendicular to the principal trajectory $\Delta\gamma = 0.0429$; resolution $R = 1000$. The luminosity value

$$\Omega = \frac{\Delta\alpha\Delta\gamma}{\pi} 100\% \tag{50}$$

calculated in accordance with the optimized parameters amounts to 0.31%.

Despite the good dispersant properties of the hyperbolic field and implementation of second-order directional focusing, the luminosity is relatively low, which is because the analyzer's electrode system does not possess the rotation axis. At the same time, the quality of focusing and dispersion value in the principal trajectory plane ($\Delta\alpha$ remains unchanged), the presence of rotation axis allows one to enlarge the collection angle in the plane perpendicular to the principal trajectory up to the value $\Delta\gamma = 2\pi$ and improve the luminosity no less than by $\pi/\Delta\gamma$ times. However, in contrast to conventional cylindrical mirror, where such a transition turns out to be possible due to the axial symmetry of the electrode system, the presence of a symmetry axis in the case of hyperbolic electrodes leads to essentially a new 3D hyperbolic field with quadratic potential distribution along all three coordinates.

2.2 3D Hyperbolic Electrostatic Fields

Potential distribution in 3D hyperbolic field. In general, the potential distribution in a 3D hyperbolic static field can be represented in the form (Gurov & Sheretov, 1984)

$$U(x, y, z) = U\left(\alpha_x x^2 + \alpha_y y^2 + \alpha_z z^2\right). \tag{51}$$

Here, U is the potential difference applied to the system of electrodes generating the potential distribution of the form of Eq. (51); α_x, α_y, α_z are certain constants defined by geometry of the electrode system that creates the hyperbolic field in question. In the approximation of a negligibly small space charge, the potential distribution [Eq. (51)] satisfies the Laplace equation, which yields the relationship between the coefficients α_i:

$$\alpha_x + \alpha_y + \alpha_z = 0 \tag{52}$$

Depending on the ratio between those coefficients, it is possible to obtain different hyperbolic field configurations. For example, if the coefficients are selected so that $\alpha_x = -\alpha_y$ and $\alpha_z = 0$, we obtain the potential distribution of a planar hyperbolic field similar to that investigated by Bredov (1959). Such a potential distribution is quadratic in the x- and y-coordinates and does not depend on the z-coordinate. The hyperboloids of revolution create an axisymmetric field, for which the relations $x_a = y_a = r_a$; $r^2 = x^2 + y^2$ and the relationship between the coefficients

$$\alpha_x = \alpha_y = \alpha_r\,,\ \alpha_z = -2\alpha_r$$

hold true. Taking this into consideration, we can represent the potential distribution [Eq. (51)] for an axisymmetric hyperboloidal electrode system in the form

$$U(r, z) = U\left(\alpha_r r^2 + \alpha_z z^2 + \alpha_0\right). \tag{53}$$

If a 3D hyperbolic field is created by the system of axisymmetric electrodes shown in Figure 7, we can represent the U value as the difference $U = \varphi_K - \varphi_T$, where φ_K is the potential of the annular electrode 1, while φ_T is mutual potential of the edge electrodes 2,3.

The coefficients α_i in the potential distribution [Eq. (53)] can be found from the boundary conditions:

$$r = r_a,\ z = 0,\ U = \varphi_K; r = 0,\ z = \pm d_a,\ U = \varphi_T. \tag{54}$$

Here, r_a and d_a are the minimal distances from the annular and end-face electrodes to the system center, respectively. Solving the system of equations that follow from the boundary conditions, we find the coefficients of the potential distribution [Eq. (53)] in the form

$$\alpha_r = \frac{1}{r_a^2 + 2d_a^2}; \alpha_z = -\frac{2}{r_a^2 + 2d_a^2}; \alpha_0 = \frac{2\varphi_K + r_a^2\varphi_T}{\left(r_a^2 + 2d_a^2\right)U} \tag{55}$$

Introduce the following notations:

$$m = \frac{\varphi_K}{U}; n = \frac{\varphi_T}{U}; \gamma = \frac{r_a}{d_a}.$$

Here, the value γ represents the geometric factor of the electrode system (i.e., the coefficient associated with geometric parameters of the system of field-generating hyperbolic electrodes); and m and n are the coefficients reflecting the features of electrical supply of the electrode system. A relationship of these parameters with geometric coefficient $n_0 = \gamma^2$ is obvious. Thus, potential distribution in the static hyperbolic field generated by a system of axisymmetric hyperbolic electrodes takes the form (Figure 7):

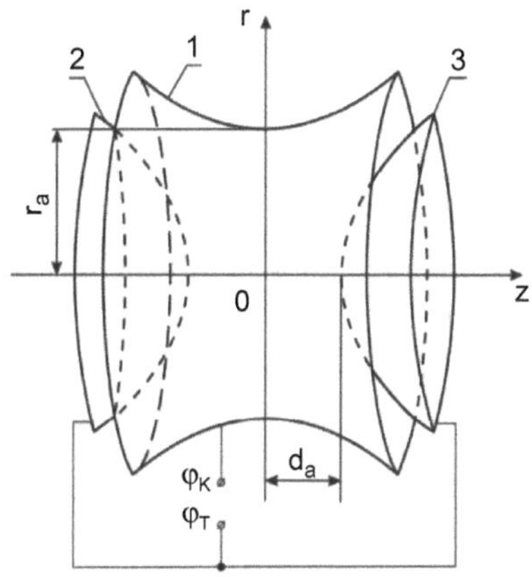

Figure 7 Hyperboloidal axisymmetric geometry of the electrode system generating a 3D electrostatic field with quadratic potential distribution: 1—annular electrode; 2, 3—end-face electrodes.

$$U(r,z) = U_0 + \frac{U}{d_a^2(2+\gamma^2)}\left(r^2 - 2z^2\right), \tag{56}$$

where $U_0 = \alpha_0 U$ is potential at the center of the hyperboloidal axisymmetric electrode system; the coefficient α_0 is determined by the expression

$$\alpha_0 = \frac{2m + \gamma^2 n}{2 + \gamma^2}. \tag{57}$$

Main types of electrode systems forming an axisymmetric hyperbolic field. Consider possible modifications of axisymmetric hyperboloidal electrode systems capable of generating 3D hyperbolic fields (Gurov, 1997). As shown previously, if we put $p_0 = 1$ into the system of equations [Eqs. (22)–(23)], the hyperboloidal electrode system will possess axial symmetry, with $x_a = y_a = r_a$ and $r^2 = x^2 + y^2$. The axisymmetric hyperboloidal electrode system (Figures 7 and 8) consists of an annular electrode (1) and two end-face electrodes—the input electrode (2) and output electrode (3).

Analytical expressions for profiles of the field-generating electrodes creating a 3D axially symmetric hyperbolic field can be obtained from the potential distribution [Eq. (56)]. By means of sequential use of the boundary conditions,

$$\varphi_T = 0,\ \varphi_K = U(r,z);\ \varphi_K = 0,\ \varphi_T = U(r,z), \tag{58}$$

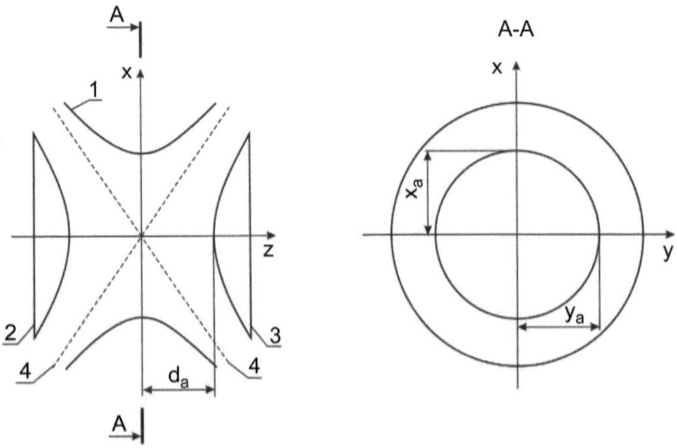

Figure 8 Hyperboloidal axisymmetric electrode system's geometry.

we obtain the profile equations for the annular electrode (labeled 1 in Figure 8):

$$r^2 = 2z^2 + r_a^2, \tag{59}$$

and the end-face electrodes (2,3),

$$2z^2 = r^2 + 2d_a^2. \tag{60}$$

In general, the equations of electrodes define one-sheet and two-sheet hyperboloids of revolution and describe the geometry of 3D axisymmetric equipotential surfaces. The hyperbolic surfaces of electrodes of the axisymmetric hyperboloidal system possess the asymptotic surfaces (labeled 4 in Figure 8). Those asymptotic surfaces are described by the following equation:

$$r^2 = 2z^2, \tag{61}$$

and represent two conical surfaces with the apexes located at the electrode system's center. The potential of the equipotential surfaces coinciding with relevant asymptotic surfaces is equal to the potential U_0 at the electrode system's center.

The equations of field-generating electrodes contain the characteristic dimensions r_a and d_a of the relevant hyperboloidal axisymmetric electrode system. To make the calculations more convenient, all the values having the dimension of length and entering into the motion equations and the equations defining the field-generating electrodes are commonly normalized by one of the characteristic dimensions of the electrode system (most often by the parameter d_a). After such a normalization, the equations of field-generating electrodes and asymptotes [Eqs. (59)–(61)] take the following forms:

Annular electrode:

$$R^2 = \gamma^2 + 2Z^2 \tag{62}$$

Face-end electrodes:

$$2Z^2 = R^2 + 2 \tag{63}$$

Asymptotic surfaces:

$$R^2 = 2Z^2 \tag{64}$$

Here,

$$R = \frac{r}{d_a};\ Z = \frac{z}{d_a};\ \gamma = \frac{r_a}{d_a}.$$

Eqs. (62) and (63), describing the field-generating electrodes of the axisymmetric hyperboloidal electrode system, are not the only possible ones that create a 3D hyperbolic field (Figure 9). Indeed, Eqs. (62) and (63) have been obtained from the potential distribution [Eq. (51)] within the assumption on the symmetry of potentials upon the end-face electrodes 2 and 3 (Figure 8). However, in general, the end-face electrodes may have different potentials; consequently, we will have different equations for the field-generating electrodes (Gurov, Trubitsyun, & Kuksa, 2015).

This makes it possible to realize the axisymmetric hyperboloidal geometry of the electrode system shown in Figure 9a. In this case, the end-face electrodes (labeled 2 and 3 in the figure) have different curvatures and d_a values ($d_{a1} < d_{a2}$). If the parameter d_{a1} of one of the end-face electrodes tends to zero, the equation describing the electrode (2) profile degenerates into Eq. (64) for the asymptote (4), and the geometry of the hyperboloidal axisymmetric system acquires a view shown in Figure 9b.

Degeneration of the second electrode (3) into a conical asymptotic electrode brings us to another type of the hyperboloidal axisymmetric electrode system capable of creating a 3D hyperbolic field (Figure 9c).

Subsequent transformation of the shape of end-face electrodes leads to the electrode system shown in Figure 9d. In this case, the axisymmetric hyperbolic field is created by two annular electrodes representing one-sheet hyperboloids of revolution. The outer and inner annular electrodes (1, 2)

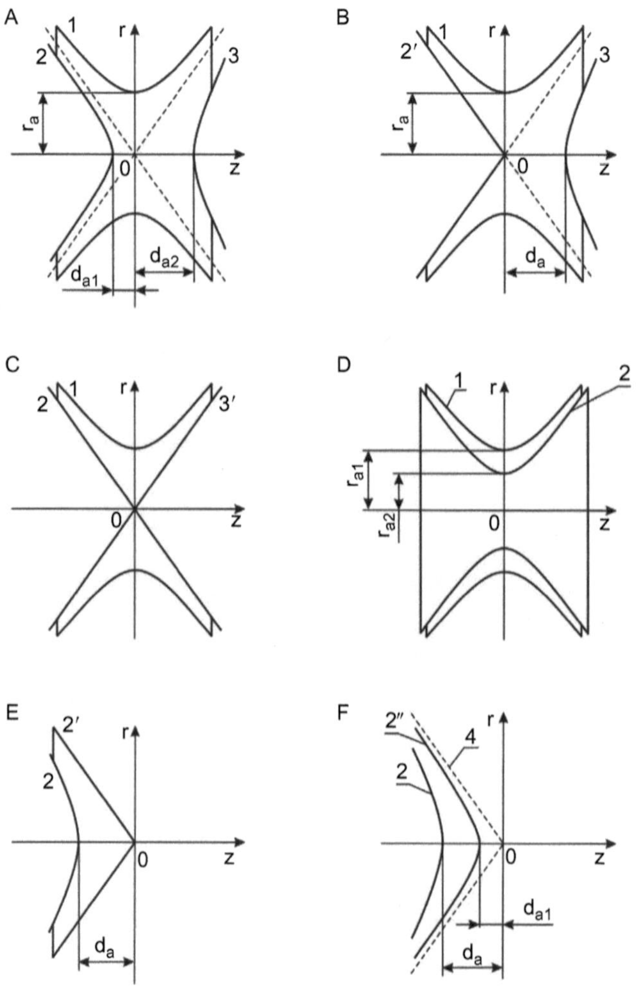

Figure 9 Various modifications of the geometry of electrode systems generating the hyperbolic axisymmetric field.

have the characteristic sizes r_{a1}, r_{a2} which represent minimal distances from the z-axis to these electrodes, respectively.

Let us derive the potential distribution in such an electrode system (Gurov & Kuksa, 2014b), assuming that the potentials of the electrodes (1, 2) are equal to φ_{a1}, φ_{a2}, respectively. From Eq. (56), we obtain the following system of equations:

$$\begin{cases} \varphi_{a1} = U\left(\alpha_r r_{a1}^2 + \alpha_0\right) \\ \varphi_{a2} = U\left(\alpha_r r_{a2}^2 + \alpha_0\right) \end{cases}. \tag{65}$$

By resolving this system with respect to $U\alpha_r$ and $U\alpha_0$, we obtain a potential distribution in the two-electrode hyperboloidal axisymmetric system (Figure 9d):

$$U(r,z)=\frac{\varphi_{a1}-\varphi_{a2}}{r_{a1}^2-r_{a2}^2}\left(r^2-2z^2\right)+\frac{\varphi_{a2}r_{a1}^2-\varphi_{a1}r_{a2}^2}{r_{a1}^2-r_{a2}^2}. \tag{66}$$

The potential distribution [Eq. (66)] brings us to the equations for the field-generating electrodes (labeled 1 and 2 in Figure 9d) by consistently setting the boundary conditions:

$$\varphi_{a1}=0,\ \varphi_{a2}=U(r,z);\ \varphi_{a2}=0,\ \varphi_{a1}=U(r,z).$$

After necessary transformations, we obtain equations for the annular electrodes (1, 2) in the following form:

$$r^2=r_{a1}^2+2z^2 \tag{67}$$

$$r^2=r_{a2}^2+2z^2. \tag{68}$$

To characterize the electrode system geometry, we introduce the geometric factor γ_0, defined as a ratio of the characteristic dimensions of annular electrodes: $\gamma_0=\frac{r_{a2}}{r_{a1}}$. When the geometric factor γ_0 tends to zero, the hyperbolic electrode (2) degenerates into two asymptotic conic surfaces that are described by the equation $r^2=2z^2$.

Finally, the axisymmetric hyperbolic field can be created in a two-electrode system (Figures 9e–f), wherein one of the electrodes represents an end-face electrode with the surface equation given by Eq. (63), while the role of the second electrode performs either the conical electrode (2′), whose generatrix is the asymptote (4) of the axisymmetric hyperboloidal electrode system (Figure 9e), or another end-face hyperbolic electrode (2″) with $d_{a1}<d_{a2}$ (Figure 9f).

Peculiarities of charged particle motion in 3D static hyperbolic fields. We consider two cases of charged particle motion in a static hyperbolic field:

- Charged particle motion in the focusing field, when the sign of the particle's charge coincides either with the sign of the potential difference $U=\varphi_K-\varphi_T$ applied to the electrode system or, in the case of zero potential of the end-face electrodes, with the sign of the annular electrode's potential
- Charged particle motion in the defocusing field, when the sign of the particle's charge is opposite to the sign of the potential difference applied to the electrode system

For separation of charged particle in energies, there is considerable interest in the charged particle motion in the focusing hyperbolic field (i.e., the mode of hyperbolic mirror). Let a charged particle enters the hyperbolic field with initial coordinates r_0 and z_0 and initial velocities $\dot{r}_0$ and $\dot{z}_0$. In this case, the charged particle trajectories can be obtained by integrating the motion equations

$$\begin{aligned} m\ddot{r} &= -qE_r \\ m\ddot{z} &= -qE_z, \end{aligned} \tag{69}$$

where

$$\begin{aligned} E_r &= \frac{\partial U(r,z)}{\partial r} = \frac{2U}{d_a^2(2+\gamma^2)}r \\ E_z &= \frac{\partial U(r,z)}{\partial z} = \frac{2U}{d_a^2(2+\gamma^2)}z. \end{aligned} \tag{70}$$

Then,

$$r = r_0 \cos\sqrt{\frac{2Uq}{md_a^2(2+\gamma^2)}}t + \frac{\dot{r}_0}{\sqrt{\frac{2Uq}{md_a^2(2+\gamma^2)}}}\sin\sqrt{\frac{2Uq}{md_a^2(2+\gamma^2)}}t; \tag{71}$$

$$z = z_0 \mathrm{ch}\sqrt{\frac{4Uq}{md_a^2(2+\gamma^2)}}t + \frac{\dot{z}_0}{\sqrt{\frac{4Uq}{md_a^2(2+\gamma^2)}}}\mathrm{sh}\sqrt{\frac{4Uq}{md_a^2(2+\gamma^2)}}t. \tag{72}$$

Let us transform Eqs. (71) and (72) by introducing the normalized time:

$$T = \sqrt{\frac{2Uq}{md_a^2(2+\gamma^2)}}t. \tag{73}$$

If the energy of a particle entering the field is equal to U_i, then the corresponding initial velocity of the particle is

$$v_i = \sqrt{2\frac{q}{m}U_i}. \tag{74}$$

For a particle that enters the hyperbolic field at the angle α to the rotation axis, the initial velocity components are

$$\begin{aligned} \dot{r}_0 &= v_i \sin\alpha \\ \dot{z}_0 &= v_i \cos\alpha. \end{aligned} \tag{75}$$

By normalizing all coordinates in the motion equations by the characteristic size d_a of the axisymmetric hyperboloidal electrode system and introducing the coefficient $P=\sqrt{\frac{U_i}{U}}$ characterizing the particle's energy, we obtain final parametric equations of the charged particle motion in the static axisymmetric focusing hyperbolic field:

$$R = R_0 \cos T + P\sqrt{2+\gamma^2} \sin\alpha \sin T; \tag{76}$$

$$Z = Z_0 \mathrm{ch}\sqrt{2}T + P\sqrt{1+0.5\gamma^2} \cos\alpha \,\mathrm{sh}\sqrt{2}T. \tag{77}$$

For the axisymmetric hyperbolic field that is formed by two annular electrodes (Figure 9d) and possesses the potential distribution [Eq. (66)], the equations of charged particle motion in the focusing field (the sign of particle's charge coincides with the sign of potential) have the form

$$\begin{cases} R = R_0 \cos kt + \left(1-\gamma_0^2\right)^{1/2} P \sin\alpha \sin kt \\ Z = Z_0 \mathrm{ch}\sqrt{2}kt + \left(\frac{1-\gamma_0^2}{2}\right)^{1/2} P \cos\alpha \,\mathrm{sh}\sqrt{2}kt \end{cases}. \tag{78}$$

In Eq. (78), R, Z, R_0, Z_0, are the particle's current and initial coordinates, respectively, and α is the angle of the particle's input to the hyperbolic field; U_i is the charged particle's energy; $P=\sqrt{\frac{U_i}{U}}$ is the energy coefficient; $\frac{q}{m}$ is the charge-to-mass ratio; and $k=\sqrt{\frac{2qU}{mr_{a1}^2\left(1-\gamma_0^2\right)}}$. All coordinates in Eq. (78) are normalized to the characteristic size r_{a1} of the annular electrode 1: $R=\frac{r}{r_{a1}}$, $Z=\frac{z}{r_{a1}}$, $R_0=\frac{r_0}{r_{a1}}$, $Z_0=\frac{z_0}{r_{a1}}$.

The trajectory analysis based on the motion equations [Eqs. (76) and (77)] shows that the charged particle trajectories in the hyperbolic field are completely determined by initial coordinates and angles of particle input into the field, as well as initial energy of particles, which is defined by the coefficient P. Suppose that a hyperbolic field is generated by the complete electrode system (Figure 10a), and the particles emitted from the point source (labeled 4 in the figure), which is located on the axis of rotation and removed from the center by the value z_i, enter the field at the angle α through the input end-face electrode (2). We can distinguish three different types of trajectories in the motion of particles entering the field under equal angles with different energies (Figure 10a):

1. Reflection of particles by the field and their return through the input end-face electrode (2; trajectory 2′)

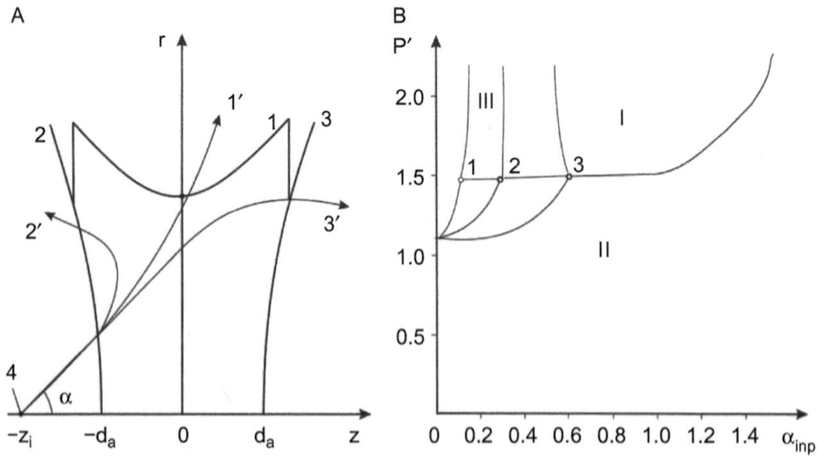

Figure 10 (A) Possible types and (B) characteristic diagram of charged particle trajectories in static axisymmetric hyperbolic field. (A) 1—annular electrode, 2—input electrode, 3—output end-face electrodes; (B) regions of particle output through the annular (I), input end-face (II) and output end-face (III) electrodes for 1—$z_i = -11d_a$; 2— $z_i = -5d_a$; 3 – $z_i = -2d_a$.

2. Passing of particles through the lens and their output through the annular electrode (1; trajectory 1′)
3. Focusing of particles by the field and their output through the output end-face electrode (3; trajectory 3′)

The reflection and return of charged particles is observed at small values of the input energies ($P < 1$). The mode of passing the particles through the annular electrode is observed at the particle energies with $P \approx 1$, while the case of $P > 1$ corresponds to focusing the particles and their output through the end-face electrode. When the angle of particle input into the hyperbolic field is changed, the energy associated with each type of trajectories also changes. However, the nature of the particle motion remains basically the same.

The characteristic diagram of different types of trajectories constructed in the coordinates $P - \alpha$ for the field of a hyperboloidal lens (Figure 10b) allows determination of the range of input energies and entry angles that correspond to the motion modes considered previously. This diagram is constructed for the case that a point source is located on the system's symmetry axis and placed at the distance $z_i = -(2 \div 11)d_a$ from the lens center; the geometric factor of the hyperboloidal electrode system is $\gamma = \sqrt{2}$, and $P' = P\sqrt{2}$. In Figure 10a, the output region number corresponds to the trajectory (and electrode) number. Analysis of the diagram shows that the widest range of input energies and angles corresponds to reflection and output of particles through the input end-face electrode. All particles with $P' < 1$ for any input angles are reflected by the field. With the input angle increase, the field reflects the particles with increasingly large energies. The particle output through the annular electrode is possible for the input angles $\alpha \geq 0.6$ at $z_i = -2d_a$ and the energies $P' \geq 1.5$. The region of particle output through the input end-face electrode is restricted by the input angles $0 < \alpha \leq 0.6$ for $z_i = -2d_a$ and the energies $P' > 1.1$. From the standpoint of gaining high-energy dispersion, of interest is the nodal point whose coordinates on the characteristic diagram at $z_i = -2d_a$ are $\alpha \approx 0.6$ and $P' \approx 1.5$. This point is common to all three regions in Figure 10b.

Changes in z_i and γ lead to a redistribution of areas of corresponding regions in the characteristic diagram (Figure 10b). In particular, a change in z_i from $-2d_a$ to $-11d_a$ leads to narrowing of region II, mainly due to narrowing of region III. Herewith, the point common for all three regions is shifted toward the region of lesser input angles. The value P' is also decreased, but to a lesser degree.

From the viewpoint of energy analyzer design, more preferred are the reflection mode—i.e., the mode of particle return through the input end-face electrode (Sheretov et al., 1987)—and the mode of focusing when the particles leave the working region through the output end-face electrode (Gurov & Sheretov, 1984). For the mode of particle output through the output end-face electrode, the dependence of the output coordinate corresponding to the plane of the output end-face electrode on the angle of particle entry into the hyperbolic field takes the form of a smooth curve with a maximum, at which the condition of first-order spatial focusing $\partial R_{out}/\partial \alpha = 0$ is satisfied (Figure 11a).

Let us call the angle satisfying the condition of first-order focusing the optimum input angle α_{1opt}. Changing the initial conditions of the input of charged particles into the axisymmetric hyperbolic field, including the input energy, point source location, and hyperboloidal electrode system's geometry, does not affect the shape of the curve $R_{out} = f(\alpha_{inp})$. Only a shift of optimum input angles occurs, herewith the increase in P or decrease in z_i leads to the increase in α_{1opt} (Figure 11a). Varying the initial conditions and geometric factor values in calculating the dependency $R_{out} = f(\alpha_{inp})$ with subsequent numerical differentiation have shown that only first-order angular focusing can be implemented if the charged particles leave the working

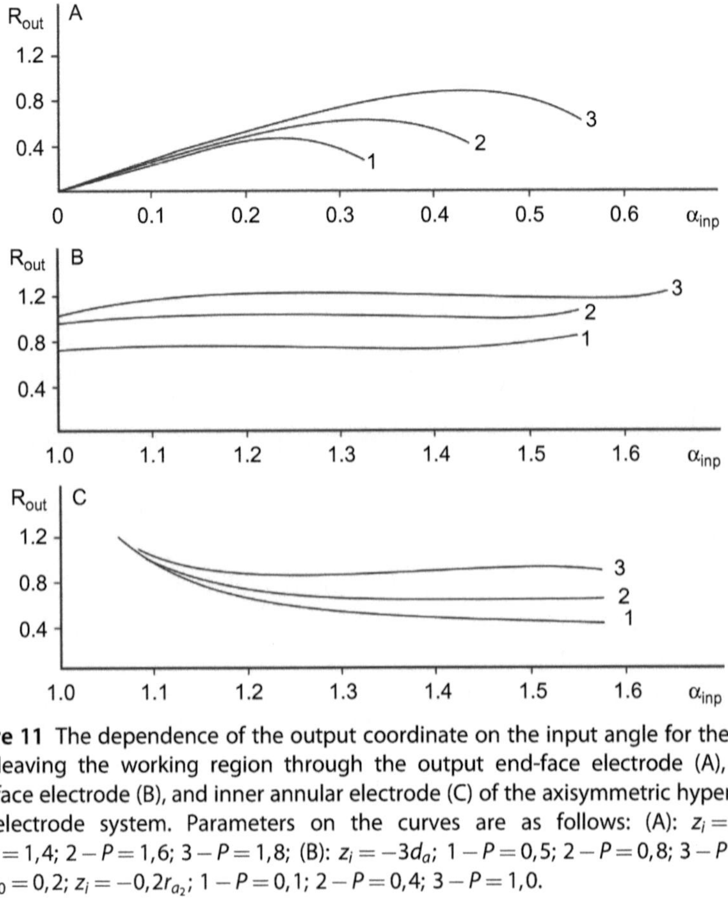

Figure 11 The dependence of the output coordinate on the input angle for the particles leaving the working region through the output end-face electrode (A), input end-face electrode (B), and inner annular electrode (C) of the axisymmetric hyperboloidal electrode system. Parameters on the curves are as follows: (A): $z_i = -3d_a$; $1 - P = 1{,}4$; $2 - P = 1{,}6$; $3 - P = 1{,}8$; (B): $z_i = -3d_a$; $1 - P = 0{,}5$; $2 - P = 0{,}8$; $3 - P = 1{,}0$; (C): $y_0 = 0{,}2$; $z_i = -0{,}2r_{a_2}$; $1 - P = 0{,}1$; $2 - P = 0{,}4$; $3 - P = 1{,}0$.

region through the output end-face electrode. The dependence of the angle α_{1opt} that ensures the mode of first-order focusing in the plane of the output end-face electrode for various conditions of inputting the particles into the axisymmetric hyperbolic field is shown in Figure 12a.

In the case of reflection mode, when the charged particles leave the working region through the input end-face electrode of the axisymmetric hyperboloidal electrode system, the dependence of the output coordinate on the angle of particle entry into the field takes the form of gently sloping curves with inflection points. The properly chosen initial conditions allow zeroing the first and second derivatives of the output coordinate as a function of the input angle of particles into the hyperbolic field (Figure 11b)—in

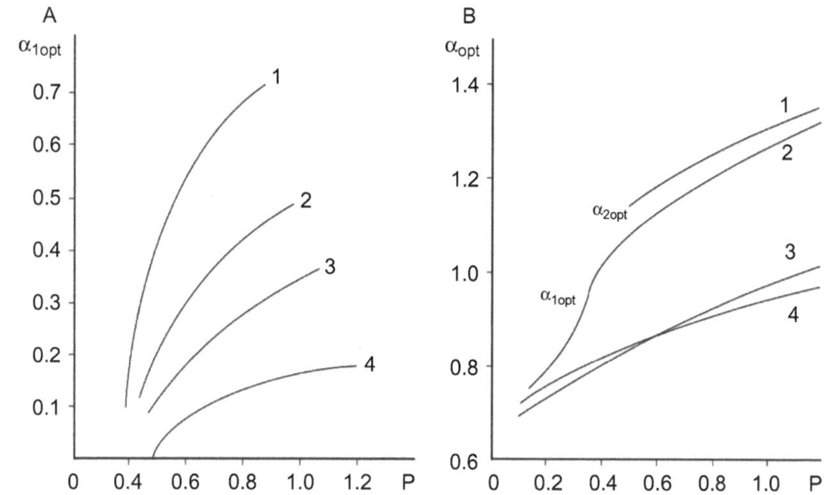

Figure 12 The dependence of the optimal input angle α_{opt} on the parameter P for the focusing (A) and return (B) modes. Parameters on the curves are as follows: (A): $\gamma = 2{,}0$; $1 - z_i = -d_a$; $2 - z_i = -2d_a$; $3 - z_i = -3d_a$; $4 - z_i = -6d_a$; (B): $1 - \gamma = 2{,}0$, $z_i = -0{,}575d_a$; $2 - \gamma = 2{,}0$, $z_i = -0{,}5d_a$; $3 - \gamma = 0{,}5$, $z_i = -2d_a$; $1 - \gamma = 0{,}2$, $z_i = -5d_a$.

other words, implementation of first- and second-order focusing in that mode. The values of angles α_{1opt} and α_{2opt} that correspond to the mode of first- and second-order focusing for different initial conditions of inputting the particles into the hyperbolic field are shown in Figure 12b. It should be noted that second-order focusing in the mode of particle reflection can be obtained in the axisymmetric hyperbolic field, provided that such a field is generated by any modification of the axisymmetric hyperboloidal electrode system shown in Figure 9, not only by means of corner electrodes as stated by Zashkvara et al. (1977).

The mode of focusing (i.e., the particle passage mode) can be implemented in a hyperbolic field generated by the axisymmetric hyperboloidal electrode system comprising two annular electrodes (Figure 9d). In this case, the input and output of particles through the inner annular electrode are determined. The dependence of the output coordinate in the plane of the inner annular electrode on the angle of particle entry into the hyperbolic field of this electrode system is more complicated (Figure 11c). For certain initial conditions, that dependence represents a monotonically descending curve (curve 1), while other initial conditions z_i, P, y_0 lead to a curve that possesses a minimum inflection point (curves 2 and 3). Such curves $[R_{out} = f(\alpha_{inp})]$ indicate the opportunity to obtain first-order and second-order angular focusing in the plane of the inner annular electrode.

Thus, the analysis of charged particle motion in axisymmetric hyperbolic fields shows that these fields exhibit a level of energy dispersion capable of spatially separating the charged particles with different energies and ensure, for certain input initial conditions, the modes of first- and second-order spatial focusing of monoenergetic flows of charged particles with the use of different ways of outputting the particles from the field.

Schematic and design parameters of axisymmetric hyperboloidal energy analyzers. The configurations of axisymmetric hyperboloidal electrode systems discussed previously allow the creation of energy analyzers with different geometries, which can be described as follows.

The particles emitted from a point source, which is located on the symmetry axis $0z$ at a distance z_i from the system center, enter the working region of the energy analyzer at the input angle α through the annular slit in the input electrode and leave the field through the annular slit of the width S in the output electrode.

By way of input and output electrodes, the end-face hyperbolic (2, 3), conical ($2'$, $3'$), and inner annular ($1'$) electrodes may be used (Figure 9). The particles with equal energies entering the working region are focused by the field into a ring of width S in the plane of output electrode within a certain range $2\Delta\alpha$ of input angles.

The average value of the input angle α_m for the particles with a given energy corresponds to the optimal angles α_{1opt} and α_{2opt} and defines the upper limit of the selection slit in the output electrode. The particles entering the field at the same angle α_m, with the energies being different from the given energy, have trajectories lying above or below the principal trajectory (for the positive and negative energy deviations, respectively). At a sufficiently small width of the selection slit S, the particles with different energies do not leave the field and hit the field-generating electrodes, which means that such a device operates as an energy analyzer. By selection of the initial entry conditions (coordinate, angle, and energy) of the particles, it is possible to implement the mode of first- and second order focusing in the plane of output selection slit. The potentials of the input and output electrodes may equal zero, and the sign of the potential of another electrode coincides with the sign of the particles analyzed.

The charged particle energy U_i on which the energy analyzer transmittance is tuned is connected with the control voltage U_c through the abovementioned energy coefficient P by the ratio $P = \sqrt{\frac{U_i}{U_c}}$. By varying the control voltage U_c on the energy analyzer electrodes in accordance with a certain law, it is possible to perform a sweep of the energy spectrum of the charged particles analyzed, since the coefficient P is constant for the design under consideration. The techniques of charged particle separation in energies commonly use a coupling coefficient (Kozlov, 1978) to relate the particle's energy to the voltage U_c on the control electrode. In this case, that coefficient is defined as $k_c = \frac{U_i}{U_c}$. Obviously, we have $k_c = P^2$.

The energy resolution and luminosity of the energy analyzer in question were determined as the design parameters. As to the energy resolution, its value was defined by the width of the instrumental function at its half-maximum, while the luminosity was calculated via the solid angle as

$$\Omega = \Delta\alpha \sin\alpha_m \times 100\%. \tag{79}$$

For the purposes of this discussion, we assume the ideality of the point source and potential distribution in the working region of the hyperboloidal axisymmetric energy analyzer. That means that the influence of the fringe fields associated with the presence of annular slits in the field-generating electrodes is disregarded.

The energy resolution and luminosity of axisymmetric hyperboloidal energy analyzers are determined by many different factors, including the geometry of the electrodes, position and width of the selection slit, point source position, energy coefficient value, and order of focusing. Herewith, as the calculations show, the effect of individual factors on the energy analyzer parameters is ambiguous and may lead to a simultaneous deterioration of one parameter and improvement of the other. Besides, the energy analyzer parameters are interrelated, and even if the dynamics of a partial impact of individual factors on the resolution and luminosity can be evaluated quite easily, the design of a real hyperboloidal axisymmetric energy analyzer, with a variety of physical and engineering requirements and limitations to be taken into consideration, represents a complicated problem of multi-objective optimization. This task would be considerably simplified if we could introduce a single efficiency parameter that is responsible for the quality of the hyperboloid energy analyzers as a whole. Obviously, such an efficiency parameter should not depend on the electrode system dimensions and selection slit width, it should only characterize the dispersion properties of the field.

Calculations of the resolution $R_{0,5}$ and luminosity Ω for the energy analyzers with the geometries considered and analysis of the contribution of different design factors, including the geometric factor γ, point source position z_i, and particle energy P, have shown that the relations

$$R_{0.5} = \frac{\nu_R(P, z_i, \gamma)}{S} \text{ and } \Omega = \nu_\Omega(P, z_i, \gamma)\sqrt{S}, \tag{80}$$

are quite well satisfied. Here, ν_R and ν_Ω represent some constants depending on P, z_i, γ. Inasmuch as the selection slit width does not define the focusing quality for charged particles, the product

$$Q = \nu_R \nu_\Omega^2 = R_{0.5}\Omega^2 \tag{81}$$

represents a constant value that is only defined by the values P, z_i, γ. With regard to the fact that the Q-value is determined only by the analysis conditions and does not depend on the width S of the selection slit, it may be used as an efficiency parameter for all types of the axisymmetric hyperboloidal energy analyzers considered (Sheretov & Gurov, 1985).

Hyperboloidal axisymmetric energy analyzer based on a complete electrode system with particle focusing. The geometry and ion-optical schematic of the energy analyzer based on a complete hyperboloidal electrode system with particle focusing is shown in Figure 13 (Gurov, 1976, Gurov & Sheretov, 1984; Sheretov & Gurov, 1980). The application of the

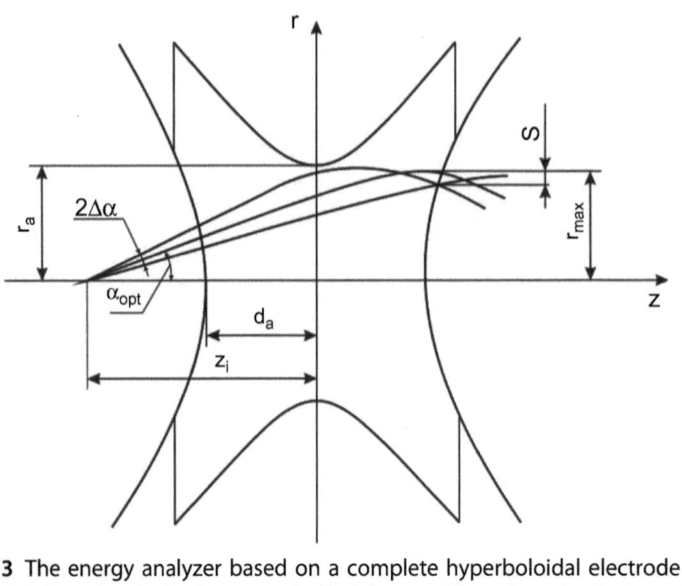

Figure 13 The energy analyzer based on a complete hyperboloidal electrode system with particle focusing (only the upper half of the particle trajectories is shown).

abovementioned technique to evaluation of the energy analyzer parameters shows the following.

The width of the selection slit strongly affects the energy resolution and luminosity of the energy analyzer. Herewith, the increase in the width of the selection slit, *ceteris paribus,* leads to an increase in the system's luminosity and a decrease in resolution (Figure 14). For example, the increase in S from 0.005 to 0.05 for the hyperboloidal electrode system with $\gamma = 2.0$ leads to an increase in luminosity from 3% to 10% and a decrease in resolution from 160 to 20. Numerical evaluation of the contribution of the selection slit width to the resolution $R_{0.5}$ and luminosity Ω of the energy analyzer based on a complete hyperboloidal electrode system for different values of the geometric factor γ, point source position z_i, and particle energy P have shown that, under the condition of first-order angular focusing in the end-face electrode's plane, the relations in Eq. (80) are well satisfied and the value of $Q = R_{0.5}\Omega^2$ can indeed act as an efficiency parameter of the energy analyzer in question.

Increasing the energy factor for fixed values of γ, z_i, and S leads to increasing the system's luminosity and energy resolution (Figure 15a). In this

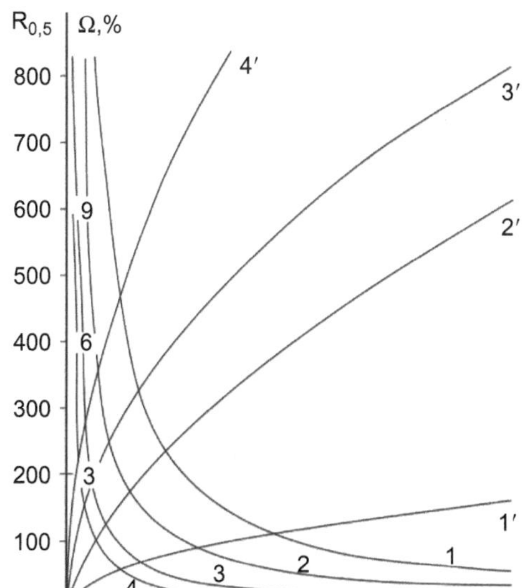

Figure 14 Dependence of the resolution $R_{0.5}$ (curves 1, 2, 3, and 4) and luminosity Ω (curves 1′,2′, 3′, and 4′) of the energy analyzer based on a complete hyperboloidal electrode system on the width S of the selection slit for different parameters of particle input into the field: $P = 1.0$; $z_i = -d_a$; $1 - \gamma = 0.5$; $2 - \gamma = 1.0$; $3 - \gamma = 2.0$; $4 - \gamma \to \infty$.

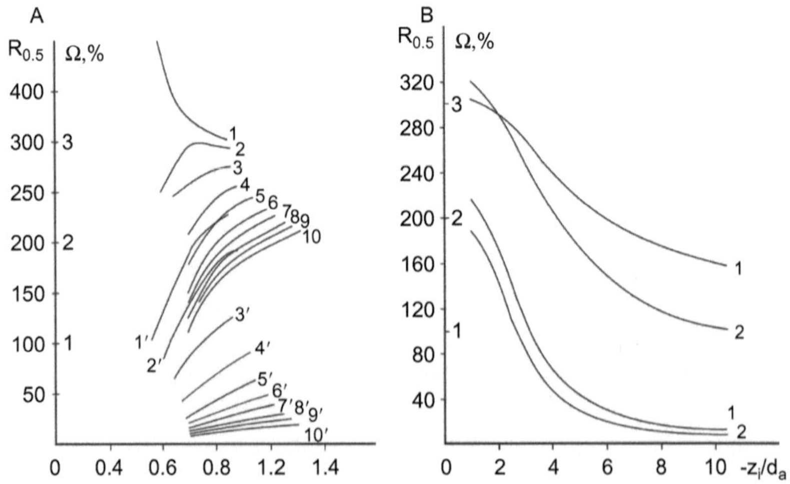

Figure 15 Effect of the energy factor P (A) and the point source location z_i (B) on the resolution $R_{0,5}$ (curves 1, 2...) and luminosity Ω (curves $1',2'$...) of the energy analyzer based on a complete hyperboloidal electrode system. Parameters on the curves are as follows: (A): $\gamma=1{,}0$; $S=0{,}005d_a$; $1-z_i=-d_a$; $2-z_i=-2d_a$; $3-z_i=-3d_a$; $4-z_i=-4d_a$; $5-z_i=-5d_a$;$6-z_i=-6d_a$;$7-z_i=-7d_a$; $8-z_i=-8d_a$;$9-z_i=-9d_a$;$10-z_i=-10d_a$; (B): $\gamma=1{,}0$; $1-P=1{,}2$; $2-P=1{,}0$.

case, the closer the point source is located to the input electrode plane, the stronger is the impact of the energy factor P on luminosity (Figure 15a, curve $1'$ and $2'$). Herewith, the energy resolution may increase in a certain range of P, and then it changes only slightly. If the point source is located in the input electrode plane, the energy factor increase results in an abrupt drop in the energy analyzer's resolution (Figure 15a, curves 1 and 2).

The effect of the point source location z_i on the energy analyzer parameters is shown in Figure 15b, which corresponds to the energy analyzer with $\gamma=1.0$, $S=0.005$, and two different energy factor values. As is seen from these curves, the parameters of the energy analyzer based on a complete hyperboloidal axisymmetric electrode system are strongly dependent on z_i; herewith, the resolution and luminosity decreased, with z_i increasing (Gurov & Sheretov, 1984).

As mentioned previously, the influence of geometric factor on resolution and luminosity is ambiguous (Figure 16). The increase in the geometric factor at a fixed width of the selection slit increases the system's luminosity and decreases its energy resolution. Herewith, two regions can be identified in these dependencies: a region of small $\gamma \leq 1$, which is characterized by high energy resolution and small luminosity, and a region of large $\gamma \geq 1$, for which

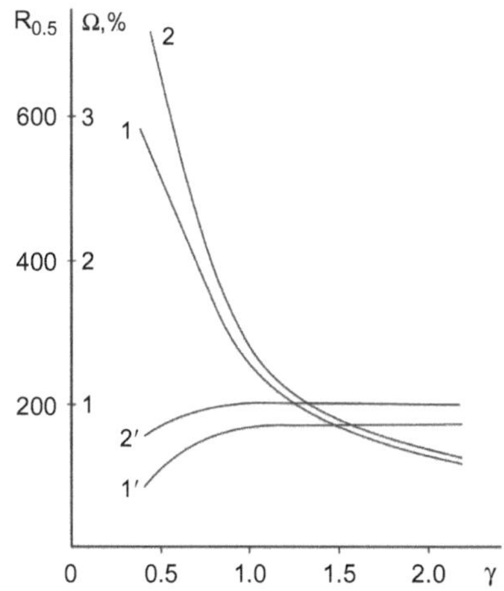

Figure 16 Effect of the geometric factor γ in a hyperboloidal electrode system on the resolution $R_{0.5}$ (curves 1, 2) and luminosity Ω (curves 1′,2′) for $z_i = -3d_a$; $1 - P = 1.0$; $2 - P = 1.1$.

the luminosity value virtually does not depend on γ, while the resolution is still quite high. Selection of the γ value allows using the hyperboloidal electrode system as a basis for creating the devices of various functional purposes, including the energy analyzers, where the main requirement is high-energy resolution, and the systems of particle collection, where the main requirement is high efficiency of particle collection (i.e., the system's luminosity).

Different versions of energy analyzers, whose parameters are determined by a set of values P, z_i, γ, can be compared to each other using the graphs displaying the system's resolution versus the luminosity. Figure 17 demonstrates such dependencies for four values of the geometric factor of a hyperboloidal axisymmetric electrode system. The calculations were performed for $P = 1{,}2$ and the value of z_i corresponding to the source position on the input electrode. The dependencies plotted in Figure 17 show that the hyperboloidal axisymmetric energy analyzer based on a complete electrode system represents a fairly high-luminosity device that ensures energy resolution of up to a few hundred at the luminosity of a few percentage points. The energy analyzer efficiency increases along with the geometric factor, and it reaches its maximum at $\rightarrow 0$, $r_a \rightarrow \infty$, which, at a fixed r_a value, corresponds to the degeneration of the hyperbolic end-face electrodes of the electrode system into the corner end-face electrodes, the surfaces of which represent the asymptotes of the hyperbolic electrodes.

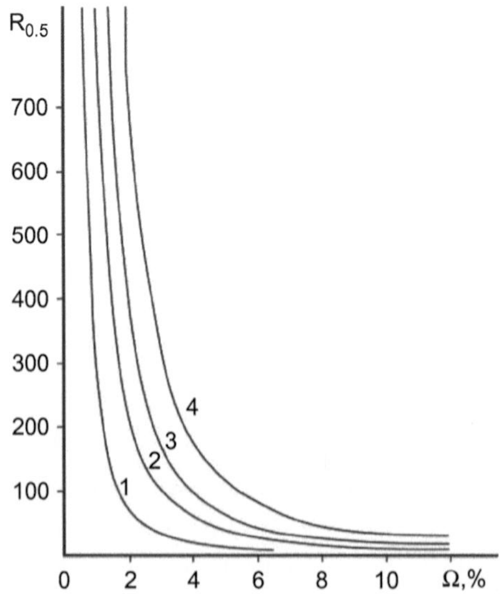

Figure 17 The dependence of the resolution $R_{0.5}$ on the luminosity Ω in the energy analyzer based on a complete electrode system for $1-\gamma=0.5; 2-\gamma=1.0; 3-\gamma=2.0; 4-\gamma\rightarrow\infty$.

Partial contribution of various factors (P, z_i, γ, S) to the resolution and luminosity is quite easy to obtain. However, the optimization of energy analyzer parameters over a whole set of the abovementioned factors and its engineering calculation represent a rather complicated problem because of contradictory influence of these factors on the output characteristic. The use of the efficiency parameter [Eq. (80)], which combines the main factors in one target function, brings that problem to the search of the maximum of the function $Q(P, z_i, \gamma)$.

It is of interest to trace the influence dynamics of the main factors on the efficiency of the energy analyzer under consideration. The results of relevant calculations are shown in Figure 18. From Figure 18a, it is clear that the strongest influence on the operation modes of the hyperboloidal energy analyzer is exerted at the point source position. In particular, a tenfold change in the parameter z_i leads to a change in Q by more than four orders of magnitude. Increasing the particle energy at the input and decreasing the parameter γ of the electrode system leads to the efficiency increase. However, such a dynamic of the geometric factor influence takes place only for the values $|z_i| > d_a$. In the range $|z_i| \approx d_a$, this condition is violated; and for $|z_i| = d_a$, when the source is positioned on the input end-face electrode, the energy analyzer's efficiency increases with increasing γ. The impact of γ on Q for

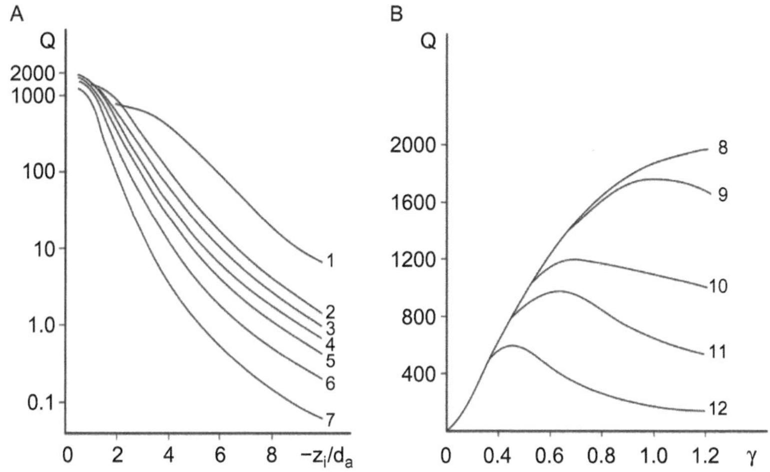

Figure 18 The dependence of the efficiency parameter Q on the distance from the point source (A) and electrode system geometry (B) for different input conditions. Parameters on the curves are as follows: 1 – $P=1{,}2$, $\gamma=0{,}5$; 2 – $P=1{,}2$, $\gamma=1{,}0$; 3 – $P=1{,}2$, $\gamma=\sqrt{2}$; 4 – $P=1{,}2$, $\gamma=2{,}0$; 5 – $P=1{,}0$, $\gamma=2{,}0$; 6 – $P=0{,}8$, $\gamma=2{,}0$; 7 – $P=0{,}6$, $\gamma=2{,}0$; 8 – $z_i=-d_a$, $P=1{,}2$; 9 – $z_i=-1{,}2d_a$, $P=1{,}2$; 10 – $z_i=-1{,}5d_a$, $P=1{,}2$; 11 – $z_i=-2d_a$, $P=1{,}2$; 12 – $z_i=-3d_a$, $P=1{,}2$.

different source positions is shown in Figure 18b. Curve (8) corresponds to the source position on the input end-face electrode, and curves (9–12) correspond to the case when the source is positioned outside the field region.

We can draw two important conclusions from the analysis of the results obtained:

- To increase the energy analyzer efficiency with the particle source positioned at the input end-face electrodes, the electrode system geometry with $\gamma\to\infty$ has to be chosen.
- Removing the particle source out of the field increases the energy analyzer efficiency, and for each z_i, an optimal value γ_{opt} exists. Herewith, for a given value γ_{opt}, maximum efficiency is achieved if the input source is positioned on the end-face electrode.

It should be noted that the complex analysis of the impact dynamics of such factors on the parameters of the hyperboloidal axisymmetric energy analyzer was only made possible with the use of the efficiency parameter, which justifies its introduction.

Hyperboloidal axisymmetric energy analyzer based on a complete electrode system with particle return. The geometry of this type of energy analyzer is shown in Figure 19 (Sheretov et al., 1987). The results of numerical

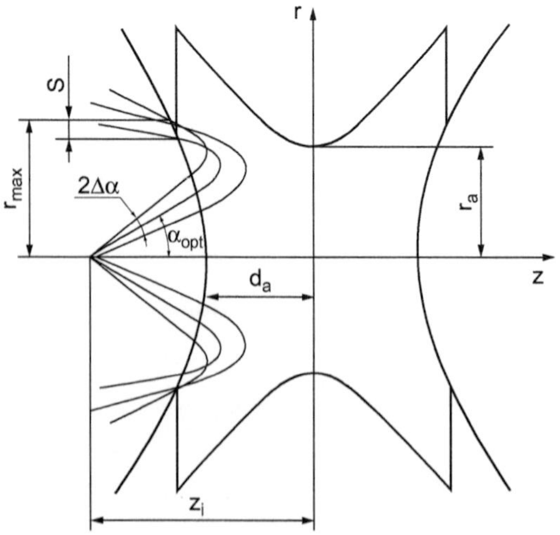

Figure 19 Energy analyzer based on a complete hyperboloidal electrode system with particle return.

evaluation of the effect of various factors on the energy analyzer parameters are presented in Figures 20–23.

The selection slit width S strongly affects the energy analyzer parameters. In particular, the increase in S from 0.001 to 0.01 at $P=0.5; z_i=1.1; \gamma=1.0$ leads to the increase in the system's luminosity from 6.5% to 22%; herewith the energy resolution falls from 800 to 75 (Figure 20a). Numerical evaluation of the slit width effect on the resolution $R_{0.5}$ and luminosity Ω of the energy analyzer in question for various values of the geometric factor γ, point source position z_i, and energy factor P has shown that the relations in Eq. (80) are quite well fulfilled, and the value $Q=R_{0.5}\Omega^2$ may be used as an efficiency parameter for this type of energy analyzer.

The particle energy contribution to the energy analyzer parameters, which may be characterized by the energy coefficient P, is also contradictory (Figure 20b), and for any initial conditions of the analyzed particles, it may lead to an increase in resolution with a simultaneous decrease in luminosity when the coefficient P is increasing. Therefore, the optimal value of P that ensures the best parameters of the energy analyzer can only be chosen using the efficiency parameter Q. The corresponding curves $Q=f(P)$ for different z_i, γ values are shown in Figure 22a. Indeed, the efficiency of the hyperboloidal energy analyzer with reflection decreases as P increases. For example, for an electrode system with $\gamma=0.5$, the increase in P from

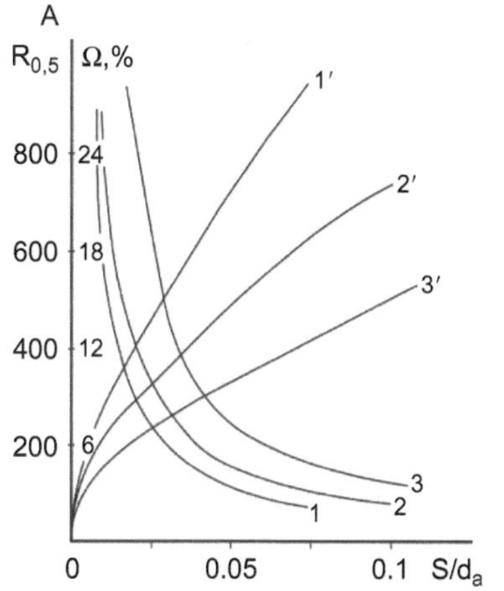

Figure 20 Effect of the selection slit width on the parameters of the energy analyzer with particle return (curves 1,2,3,4 – resolution $R_{0,5}$; 1′,2′,3′,4′ – luminosity Ω). Parameters on the curves are as follows: (A): $P=0,5$; $1-\gamma=\sqrt{2}$, $z_i=-0,8d_a$; $2-\gamma=1,0$, $z_i=-1,1d_a$; $3-\gamma=0,5$, $z_i=-2,15d_a$; (B): $1-\gamma=\sqrt{2}$, $z_i=-0,8d_a$; $2-\gamma=1,0$, $z_i=-1,1d_a$; $3-\gamma=0,2, z_i=-5,4d_a$; $4-\gamma=0,2, z_i=-5d_a$.

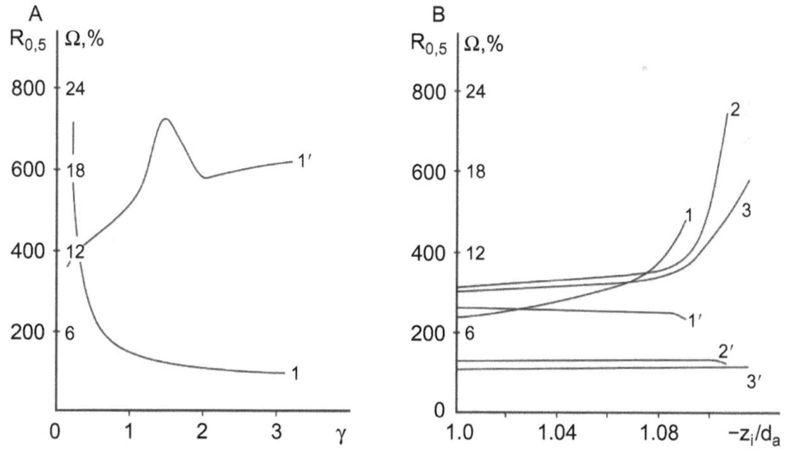

Figure 21 Effect of the geometric factor (A) and the point source position (B) on the parameters of the energy analyzer with particle return (curves 1,2,3 – resolution $R_{0,5}$; 1′,2′,3′ – luminosity Ω). Parameters on the curves are as follows: (A): $1-P=0,5$, $z_i=-1,1d_a$; (B): $P=0,5$; $1-\gamma=0,5$; $2-\gamma=\sqrt{2}$; $3-\gamma=2,0$.

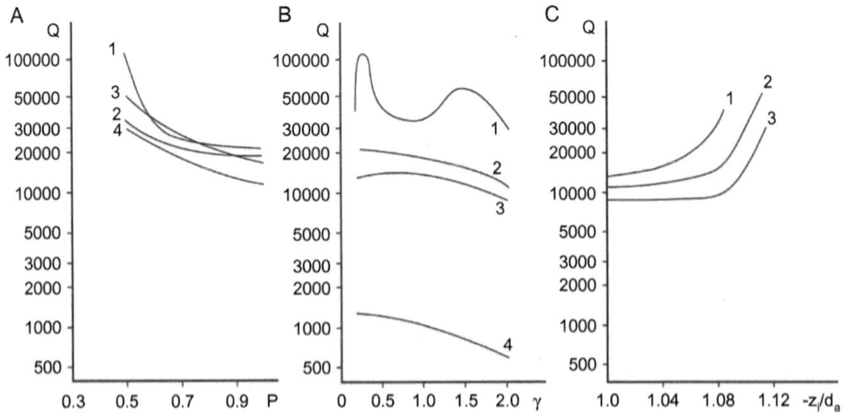

Figure 22 Effect of the energy factor P (A), geometric factor γ (B), and the point source position (C) on the efficiency parameter Q in the energy analyzer with particle return. Parameters on the curves are as follows: (A): $z_i = -1,1d_a$; $1-\gamma = 0,5$; $2-\gamma = 1,0$; $3-\gamma = \sqrt{2}$; $4-\gamma = 2,0$; (B): $z_i = -1,1d_a$;$1-P = 0,5$; $2-P = 0,8$; $3-P = 0,9$; $4-P = 1,0$; (C): $P = 0,5$; $1-\gamma = 0,5$; $2-\gamma = \sqrt{2}$; $3-\gamma = 2,0$.

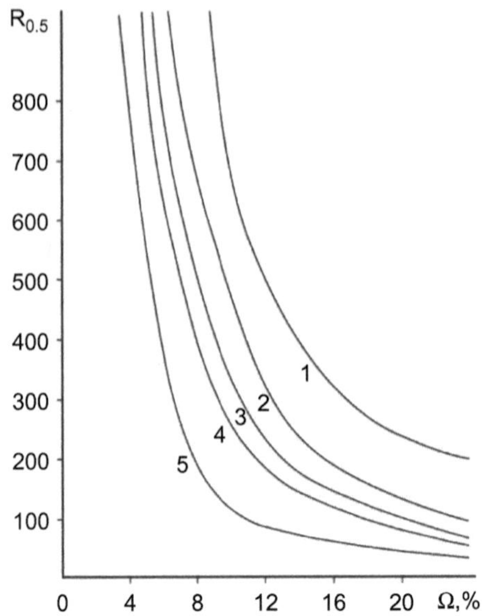

Figure 23 The dependence of the resolution $R_{0.5}$ on the luminosity in the energy analyzer with particle return for different conditions of particle input: $1-\gamma = 0.2$, $z_i = -5.4d_a$; $2-\gamma = \sqrt{2}$, $z_i = -0.799d_a$; $3-\gamma = 1.0$, $z_i = -1.1d_a$; $4-\gamma = 0.5$, $z_i = -2.15d_a$; $5-\gamma = 0.2$, $z_i = -5d_a$.

0.5 to 0.9 reduces the value Q from to 1.1×10^5 to 2.2×10^4 (i.e., almost by an order of magnitude).

The hyperboloid geometry of the electrode system, which is characterized by the geometric factor γ, also produces an ambiguous effect on the energy analyzer parameters. The increase in γ commonly increases the luminosity and decreases the energy resolution (Figure 21a). The analysis based on the use of the efficiency parameter shows (Figure 22b) that the optimum geometry for the energy analyzer with particle return represents the electrode system with $\gamma\leq1.5$; herewith, for certain input initial conditions, the maximum efficiency is reached at $\gamma=0.2$. On the whole, it may be stated that the increase in γ leads to the decrease in efficiency of the energy analyzer with particle return (Figure 22b).

This analysis of the peculiarities of charged particle motion in the hyperbolic field in the course of particle reflection and return has shown that, under certain input initial conditions, it is possible to obtain the focusing modes of not only the first order, but also the second order. For given values γ and P, the mode of second-order focusing may be implemented by selecting the value of Z_i. The impact of the point source position on the parameters of the energy analyzer with particle return is shown in Figure 21b. The calculations show that the second-order focusing is accompanied by a sharp increase in the system luminosity and a certain amount of decrease in the resolution, and is achieved at a fixed value of S, with the parameters z_i, P, and γ interconnected by the relation $z_i=-\frac{f(P)}{\gamma}$. It should be noted that $f(P)=1.0875$ at $P=0.5$.

The relation $z_i\gamma=1$ corresponds to the case when the source is positioned on the input electrode of the analyzer. Placing the source outside the field yields a sharp increase in luminosity and a definite decrease in resolution (Figure 21b), although in general, the efficiency of energy analysis increases as z_i increases, which enhances the efficiency factor Q by 4–5 times for a given γ (Figure 22c) and almost by 10 times in the case of optimal electrode system geometry (Figure 22b, curves 1, 3).

The graphs of the resolution versus luminosity for the modes of first- and second-order focusing as applied to the energy analyzer based on a complete hyperboloidal electrode system with particle return are presented in Figure 23. These graphs show that the implementation of the mode of second-order focusing allows one to increase the device luminosity by 2.5–3 times with the same electrode system geometry, energy coefficient P, and resolution. In particular, it is possible to obtain a resolution equal to 100 at luminosity values up to 25%–30% (Sheretov, et al., 1987). It should be noted that the best energy analyzers based on a cylindrical mirror provide a luminosity of 7%–8% at the same resolution (Gorelik et al., 1976).

Hyperboloidal axisymmetric energy analyzer with a single conic end-face electrode (focusing mode). The energy analyzer of this type is shown in Figure 24. The results of relevant calculations (Gurov & Kuksa, 2013) in the focusing mode are presented in Figures 25–27.

The influence of the selection slit width on the energy analyzer parameters (Figure 25a) is similar to that of the energy analyzers already considered. The calculations performed for different input conditions have shown that the operation efficiency of this energy analyzer may be characterized by the efficiency parameter $Q = R_{0.5}\Omega^2$ defined by a set of values (p, z_i, γ).

Increasing the energy factor P leads to the increase in the energy analyzer resolution and produces a weak effect on the luminosity, which is determined by the values γ and z_i (Figure 25b). Decreasing the luminosity as parameter P increases is observed only if the particle source is positioned on the input electrode ($z_i = 0$). The dependence of the efficiency parameter on the energy factor is also ambiguous (Figure 27a). If the source is positioned on the input electrode ($z_i = 0$) and $\gamma > 0.5$, the energy analyzer efficiency falls as the parameter P increases. At $\gamma = 0.5$ and $z_i = 0$, the dependence $Q = f(P)$ possesses a maximum at $P = 0.3$. At $\gamma = 0.5$ and $z_i = 0$,

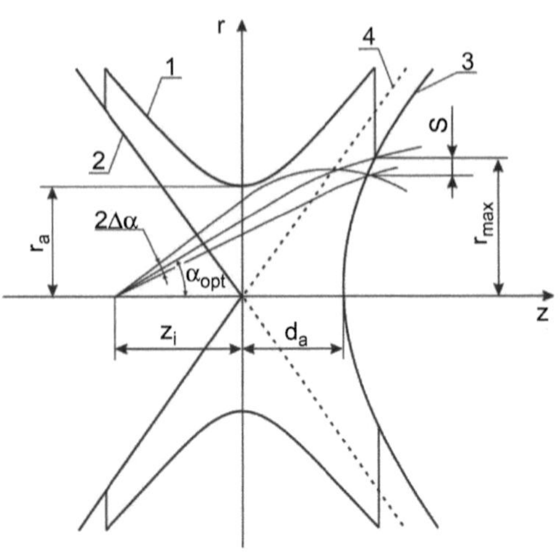

Figure 24 Hyperboloidal axisymmetric energy analyzer with a single conical end-face electrode (charged particle trajectories in the upper half-plane are shown).

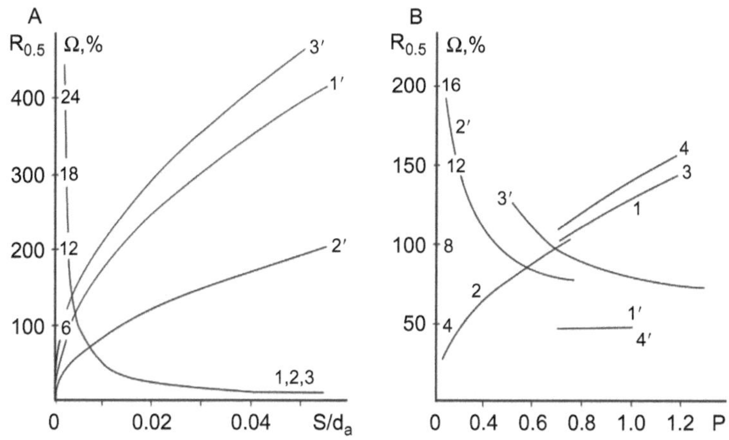

Figure 25 Effect of the selection slit width S (A) and the energy factor P (B) on the parameters of the energy analyzer with a single conical end-face electrode (curves 1,2,3,4 – energy resolution $R_{0,5}$; 1′,2′,3′,4′ – luminosity Ω). Parameters on the curves are as follows: (A): $P = 1{,}0$; $1 - \gamma = 1{,}0$, $z_i = 0$; $2 - \gamma = 1{,}0$, $z_i = -d_a$; $3 - \gamma = 0{,}5$, $z_i = -0$; (B): $1 - \gamma = 2{,}0$, $z_i = -d_a$; $2 - \gamma = 2{,}0$, $z_i = -0$; $3 - \gamma = 1{,}0$, $z_i = -0$; $4 - \gamma = 1{,}0$, $z_i = -d_a$.

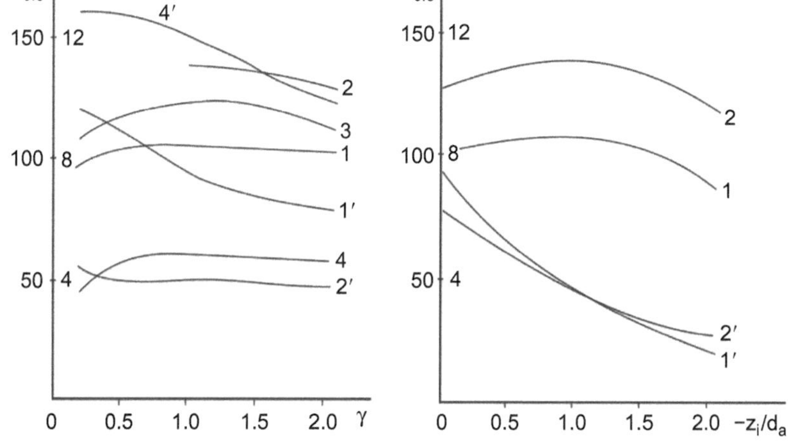

Figure 26 Effect of the geometric factor γ (A) and the point source position z_i (B) on the parameters of the energy analyzer with a single end-face electrode (curves 1,2,3,4 – energy resolution $R_{0,5}$; 1′,2′,3′,4′ – luminosity Ω). Parameters on the curves are as follows: (A): $1 - P = 1{,}0$, $z_i = 0$; $2 - P = 1{,}6$, $z_i = -d_a$; $3 - P = 1{,}2$, $z_i = -d_a$; $4 - P = 0{,}3$, $z_i = -0$; (B): $1 - \gamma = 1{,}0$, $P = 1{,}0$; $2 - \gamma = 1{,}0$, $P = 1{,}6$.

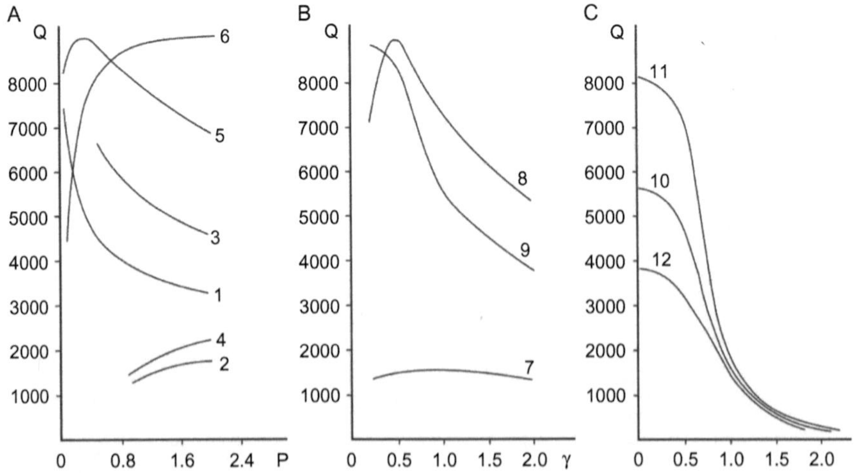

Figure 27 Effect of the energy factor P (A), geometric factor (B), and the point source position z_i (C) on the efficiency parameter Q of the energy analyzer with a single conical end-face electrode. Parameters on the curves are as follows: (A): 1 $-\gamma=2{,}0$, $z_i=0$; 2 $-\gamma=2{,}0$, $z_i=-d_a$; 3 $-\gamma=1{,}0$, $z_i=0$; 4 $-\gamma=1{,}0$, $z_i=-d_a$; 5 $-\gamma=0{,}5$, $z_i=0$; 6 $-\gamma=0{,}2$, $z_i=0$; (B): 7 $-P=2{,}0$, $z_i=-d_a$; 8 $-P=0{,}3$, $z_i=0$; 9 $-P=1{,}0$, $z_i=0$; (C): 10 $-\gamma=1{,}0$, $P=1{,}0$; 11 $-\gamma=0{,}5$, $P=1{,}0$; 12 $-\gamma=2{,}0$, $P=1{,}0$.

the energy analyzer efficiency increases as P increases and changes slightly when $P>1.2$. In the case of particle input from the field-free region, the energy analyzer efficiency always increases as P increases.

The influence of the electrode system geometry on the energy analyzer parameters, as well as its efficiency, are shown in Figures 26b and 27b. It can be seen that if the charged particle source is placed outside the field, the optimal value of γ is $\gamma=1$ and the function $Q=f(\gamma)$ has a maximum. If the source is positioned on the input electrode, all the energy analyzer parameters strongly depend on γ. For $P<1.0$, optimal γ values lie in the region $\gamma<0.5$; in this case, the energy analyzer efficiency is 2.5–3 times higher than for $\gamma=2.0$.

Increasing the point source distance to the system's center reduces the energy analyzer luminosity (Figure 26b). The resolution dependence on the distance z_i has a maximum of $z_i=-1.2d_a$. In general, the energy analyzer efficiency falls as z_i increases; herewith the change in z_i from 0 to $-2d_a$ leads to the reduction of Q from 8.2×10^3 to 2.5×10^2; i.e., more than by 30 times (Figure 27 c).

The resolution of a hyperboloidal energy analyzer with a conical input end-face electrode versus the luminosity for different input conditions is

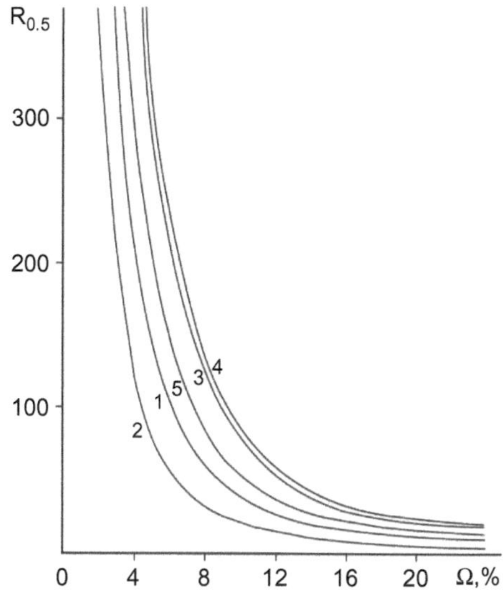

Figure 28 The dependence of resolution $R_{0.5}$ on luminosity Ω in the energy analyzer with a conical end-face electrode under various input conditions: $1-\gamma=2.0$, $P=1.0$, $z_i=0$; $2-\gamma=2.0$, $P=1.0$, $z_i=-0.5d_{ai}$; $3-\gamma=0.5$, $P=1.0$, $z_i=0$; $4-\gamma=0.2$, $P=1.0$, $z_i=0$; $5-\gamma=\sqrt{2}$, $P=1.0$, $z_i=0$.

shown in Figure 28. These curves show that the highest luminosity among devices of this type possesses the energy analyzer with $=0.2$, having the input source of particles positioned on the input electrode $(z_i=0)$. Such a device ensures the luminosity about 10% at the resolution of 100, which is greater than that of the best energy analyzers of the cylindrical mirror type (Gorelik et al., 1976).

Hyperboloidal axisymmetric energy analyzer with two conical end-face electrodes (focusing mode). The energy analyzer of this type is shown in Figure 29. The results of numerical analysis of the impact of various factors on the parameters of such a device are shown in Figures 30 and 31 (Sheretov, Zenkin, Safonov, Gurov & Sakhanova, 1985; Gurov & Kuksa, 2014a).

The effect of the selection slit's width in the case under consideration is analogous to that in the energy analyzer already considered (Figure 30a). The calculations performed for different values of the selection slit's width and initial conditions of particle input into the analyzer's field have shown that the hyperboloidal energy analyzer with angled end-face electrodes can also be characterized by the efficiency parameter $Q=R_{0.5}\Omega^2$ [Eq. (80)].

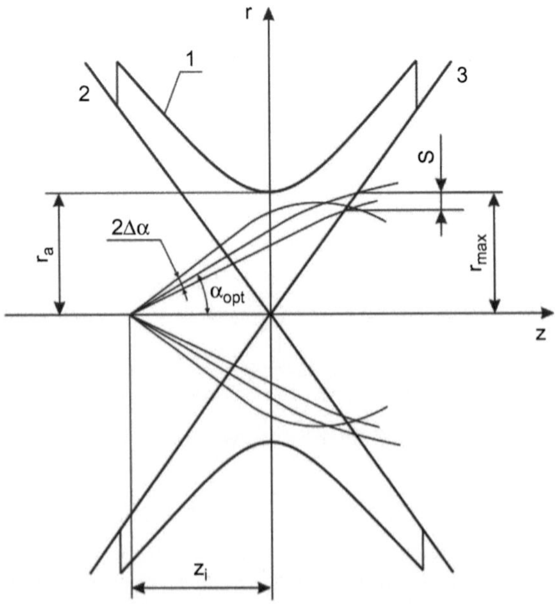

Figure 29 Hyperboloidal axisymmetric energy analyzer with two conical end-face electrodes.

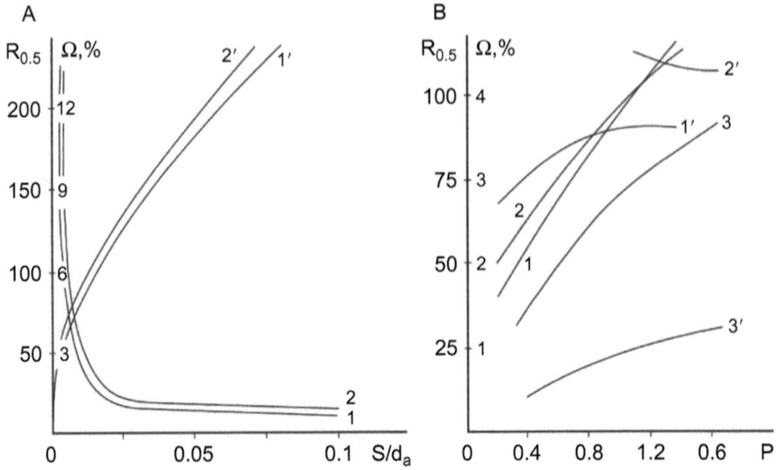

Figure 30 Effect of the selection slit's width S (A) and energy factor P (B) on the parameters of the energy analyzer with two conical end-face electrodes (curves 1,2,3—resolution $R_{0.5}$; 1′, 2′, 3′—luminosity Ω). (A) $1-\gamma=1.0, P=1.0, z_i=-d_a$; $2-\gamma=1.0, P=1.3, z_i=-d_a$; (B) $\gamma=1.0$; $1-z_i=-d_a$; $2-z_i=-0.5d_a$; $3-z_i=-3d_a$.

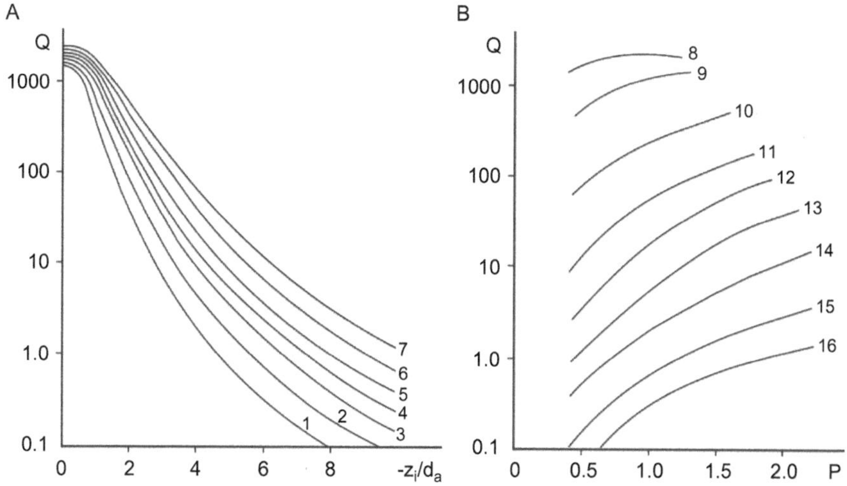

Figure 31 Effect of the point source position z_i (A) and the energy factor P (B) on the efficiency parameter Q in the energy analyzer with two conical end-face electrodes. Parameters on the curves are as follows: (A): $1-P=0{,}4$; $2-P=0{,}6$; $3-P=0{,}8$; $4-P=1{,}0$; $5-P=1{,}2$; $6-P=1{,}6$; $7-P=2{,}0$; (B): $8-z_i=-0{,}5d_a$; $9-z_i=-d_a$; $10-z_i=-2d_a$; $11-z_i=-3d_a$; $12-z_i=-4d_a$; $13-z_i=-5d_a$; $14-z_i=-6d_a$;$15-z_i=-8d_a$; $16-z_i=-10d_a$.

Increasing the energy factor P, along with reducing the distance z_i to the point source, leads to the increase in the system's resolution and luminosity (Figure 30b) along with the increase in the energy analyzer's efficiency (Figure 31). In the energy analyzers of this type, the efficiency parameter Q attains its maximum when the source is positioned on the input electrode. However, at small values of z_i, the influence of P is also small, so that the efficiency parameter remains invariable in a wide range of the energy factor values.

Hyperboloidal axisymmetric energy analyzer with two annular electrodes. The geometry of this energy analyzer is displayed in Figure 32, while the relevant calculation results are presented in Figures 33–35 (Gurov, Sheretov, & Sakhanova, 1985; Gurov & Kuksa, 2014b).

Increasing the selection slit width increases the energy analyzer luminosity and decreases its resolution. This situation is observed for all conditions of particle input into the analyzer (Figure 33a). As the calculations show, the efficiency of the energy analyzer with two annular electrodes can also be characterized by the efficiency parameter $Q=R_{0.5}\Omega^2$.

The influence of the energy factor P on the energy analyzer parameters is unambiguous—the resolution, luminosity (Figure 33b), and efficiency Q

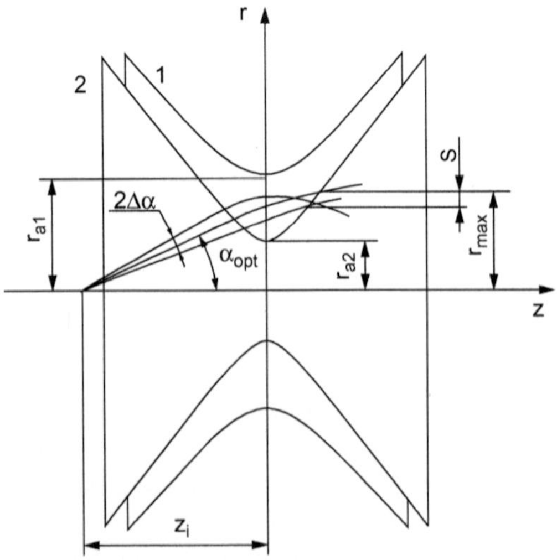

Figure 32 Hyperboloidal axisymmetric energy analyzer with two annular electrodes.

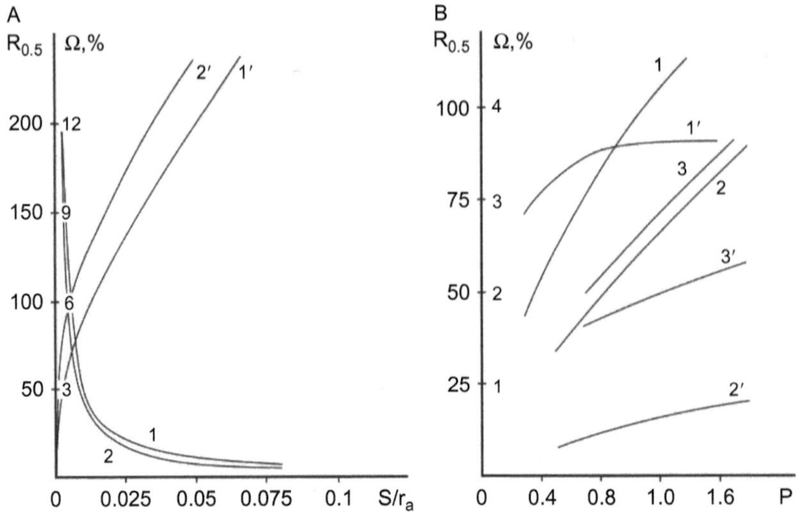

Figure 33 Effect of the selection slit's width S (A) and the energy factor P (B) on the parameters of the energy analyzer with two annular electrodes (curves 1,2,3 – resolution $R_{0,5}$; 1′,2′,3′ – luminosity Ω). Parameters on the curves are as follows: (A): $1-y_0=0{,}2$, $P=1{,}1$, $z_i=-0{,}4r_{a1}$; $2-y_0=0{,}4$, $P=1{,}0, z_i=-0{,}4r_{a1}$; (B): $1-y_0=0{,}1$, $z_i=-r_{a1}$; $2-y_0=0{,}2$, $z_i=-4r_{a1}$; $3-y_0=0{,}4$, $z_i=-2r_{a1}$.

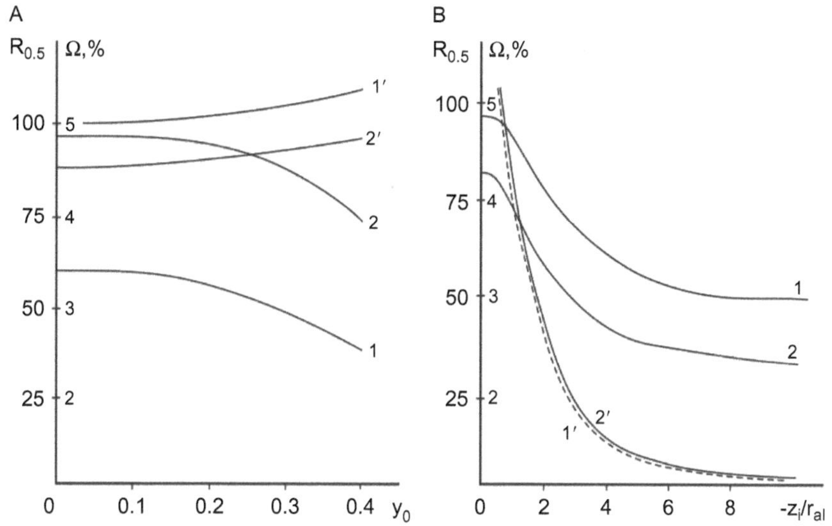

Figure 34 Effect of the geometric factor y_0 (A) and the point source position z_i (B) on the parameters of the energy analyzer with two annular electrodes (curves 1,2 – resolution $R_{0,5}$; 1′,2′ – luminosity Ω). Parameters on the curves are as follows: (A): $1-z_i=-0{,}5r_{a1}$, $P=0{,}4$; $2-z_i=-0{,}5r_{a1}$, $P=1{,}0$; (B): $1-y_0=0{,}1$, $P=1{,}0$; $2-y_0=0{,}4$, $P=1{,}0$.

(Figure 35a) increase as P increases; herewith the increase rate falls when the particle source comes close to the system's center.

In contrast to other types of hyperboloidal axisymmetric energy analyzers, the energy analyzer with two hyperbolic annular electrodes has a specific geometric parameter $y_0=r_{a2}/r_{a1}$, where r_{a1},r_{a2} are the characteristic dimensions of external and interior annular electrodes, respectively. If $y_0=0$, the interior annular electrode degenerates into the angled electrodes representing asymptotes of the hyperboloidal electrode system. As can be seen from Figure 34a, the increase in y_0 increases the system's luminosity and decreases the resolution; however, in general, the energy analyzer efficiency Q does not change and starts to decrease slightly only in the region $y_0\geq 0{,}4$ (Figure 35b).

In the case of a hyperboloidal axisymmetric energy analyzer with two annular electrodes, the increase in the distance of the point source z_i leads to a significant decrease in the resolution and luminosity (Figure 34b). Varying z_i from -1 to -10 reduces the efficiency parameter by more than four orders of magnitude (Figure 35b).

The resolution behavior as a function of the luminosity for different geometries of the energy analyzer under consideration (Figure 36) indicates the possibility of designing a hyperboloidal axisymmetric energy analyzer

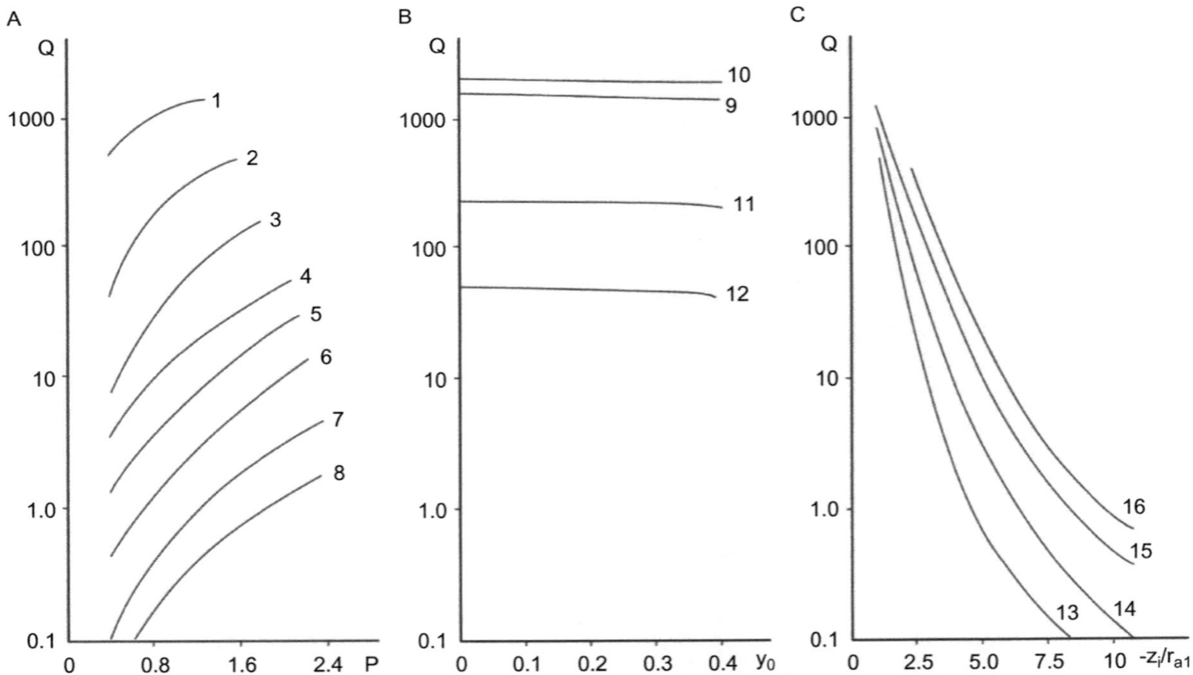

Figure 35 Effect of the energy factor P (A), geometrical factor γ (B), and the point source position z_i (C) on the efficiency parameter Q in the energy analyzer with two annular electrodes. Parameters on the curves are as follows: (A): $y_0=0,1$; $1-z_i=-r_{a1}$; $2-z_i=-2r_{a1}$; $3-z_i=-3r_{a1}$; $4-z_i=-4r_{a1}$; $5-z_i=-5r_{a1}$; $6-z_i=-6r_{a1}$; $7-z_i=-7r_{a1}$; $8-z_i=-10r_{a1}$; (B): $9-P=0,4$, $z_i=-0,5r_{a1}$; $10-P=1,0$, $z_i=-0,5r_{a1}$; $11-P=1,0$, $z_i=-2r_{a1}$; $12-P=1,0$, $z_i=-3r_{a1}$; (C): $y_0=0,1$; $13-P=0,4$; $14-P=0,8$; $15-P=1,4$; $16-P=1,8$.

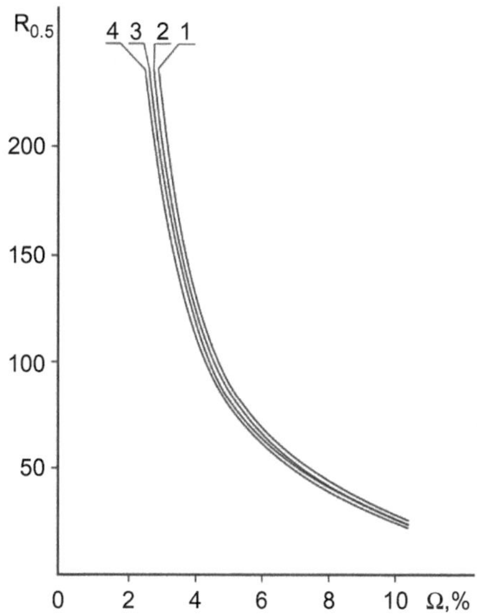

Figure 36 The dependence of the resolution $R_{0.5}$ on the luminosity Ω in the energy analyzer with two annular electrodes for various conditions of particle input: $P = 1.2$; $z_i = -0.5r_{a1}$; $1 - y_0 = 0.1$; $2 - y_0 = 0.2$; $3 - y_0 = 0.3$; $4 - y_0 = 0.4$.

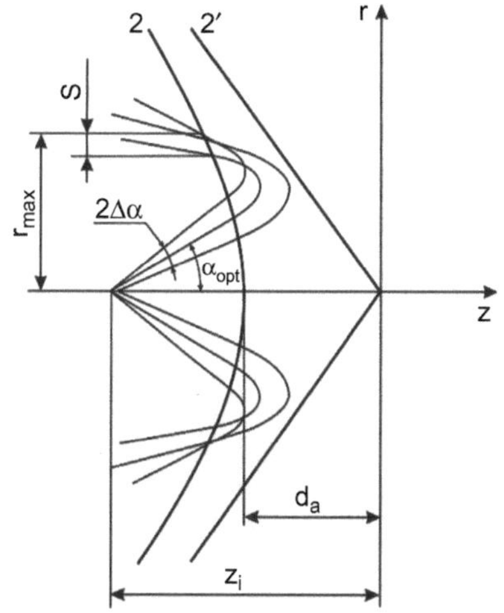

Figure 37 Two-electrode hyperboloidal axisymmetric energy analyzer with particle return.

with two annular electrodes that are capable of providing a resolution of 100 at luminosities up to 4.5%.

Two-electrode hyperboloidal axisymmetric energy analyzer with particle return. This type of energy analyzer (Figure 37) (Gurov, 1977) is similar to

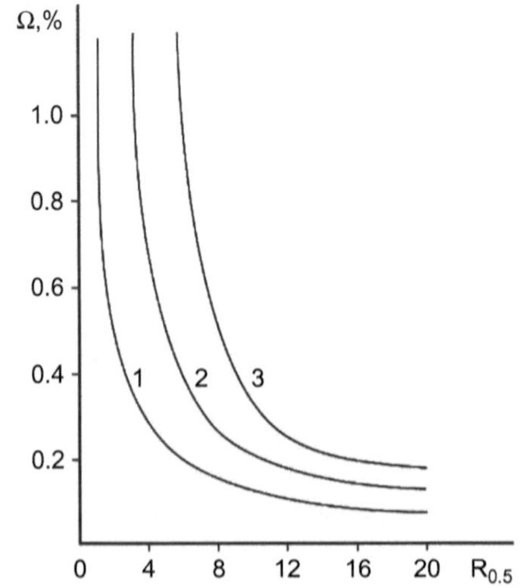

Figure 38 The dependence of the luminosity Ω on the resolution $R_{0.5}$ in the two-electrode energy analyzer with particle return for $z_i = -2d_a$; $1 = P = 0.3$; $2 - P = 0.4$; $3 - P = 0.5$.

the energy analyzer based on a complete electrode system with particle return (Figure 38). The difference is only in a simpler design, which was a reason to investigate the possibility of using such an electrode system to create an energy analyzer capable of forming an output flow with particle radial velocity equal to zero. The calculation method does not differ from that described here, except that the optimum angle is chosen not from the focusing condition, but rather meets the condition of zero radial velocity. The luminosity dependence on resolution for such a device is shown in Figure 38. Analysis of the results shows that the system's luminosity does not exceed 0.2% at a resolution of 20.

Implementation of the conditions of first- or second-order focusing in this two-electrode system ensures the design of a small hyperboloidal energy analyzer with particle return (reflection), the parameters of which were described earlier in this chapter.

Hyperboloidal axisymmetric energy analyzer with selection slits positioned in the drift space. Let us analyze the dependencies of the output coordinates and velocities on the different conditions of charged particle input into the electrostatic field of a hyperboloidal axisymmetric electrode system Gurov, Trubitsyn, Grachev, and Kuksa (2015). Assume that the particles enter the hyperbolic field through the input electrode from a point source located on the system's symmetry axis. The dependencies of the output

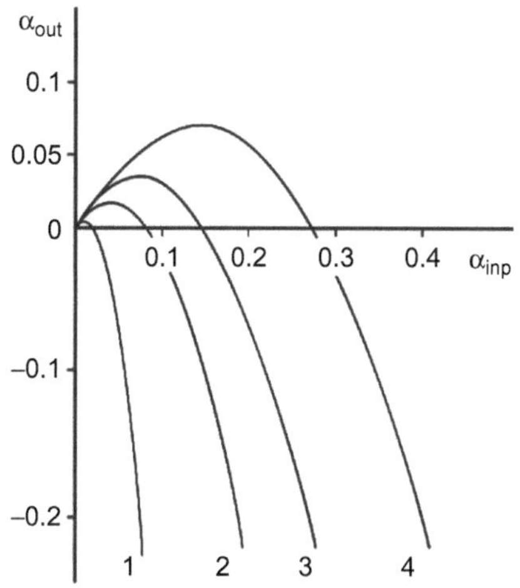

Figure 39 The dependence of the angle α_{out} of the particle output through the end-face electrode for a complete hyperboloidal electrode system with $\gamma = 1.0$ for $1 - P = 0.04$, $z_i = -3d_a$; $2 - P = 0.6$, $z_i = -3d_a$; $3 - P = 0.8$, $z_i = -2d_a$; $4 - P = 1.6$, $z_i = -2d_a$.

particle angles α_{out} on the input angles α_{inp} for a complete hyperboloidal electrode system at different z_i and P are shown in Figure 39. It can be seen that the change in the initial input conditions allows for obtaining both the divergent ($\alpha_{out} > 0$) and convergent ($\alpha_{out} < 0$) particle flows. However, the creation of efficient energy analyzers with slightly diverging particle flows ($\alpha_{out} \cong 0$) is difficult because the condition of first-order spatial focusing in the output electrode plane corresponds to the values of output angles $\alpha_{out} \ll 0$.

To study the possibility of increasing the focal length of the axisymmetric hyperboloidal energy analyzers with maintaining high analytical characteristics (energy resolution and luminosity), we have to consider the possibility of spatial focusing of particles in the field-free region at the output from the hyperbolic field. The position of the electrode with selection slit in the drift space is not known beforehand; therefore, we must use the following procedure to determine the first- or second-order conditions in the drift space.

Let us formulate the basic equation of a charged particle trajectory in the form

$$r = \psi(z, P, z_i, \alpha), \tag{82}$$

where P, Z_i and α are the initial conditions of particle motion in the field. For an adjacent trajectory having the same values of P and Z_i and a small increment of the input angle $\Delta\alpha$ (i.e., $\alpha' = \alpha + \Delta\alpha$), this equation can be written as

$$r_2 = \psi(z, P, z_i, \alpha') = \psi(z, P, z_i, \alpha') + \frac{1}{1!}\Delta\alpha\frac{\partial\psi}{\partial\alpha} + \frac{1}{2!}\frac{\partial^2\psi}{\partial\alpha^2}\Delta\alpha^2 + \ldots. \quad (83)$$

If these trajectories intersect at some point z_i, we have $r_1 = r_2$ at that point, and

$$-\frac{\partial\psi}{\partial\alpha} = \frac{1}{2!}(\Delta\alpha)\frac{\partial^2\psi}{\partial\alpha^2} + \frac{1}{3!}(\Delta\alpha)^2\frac{\partial^3\psi}{\partial\alpha^3} + \ldots. \quad (84)$$

The value $\frac{\partial\psi}{\partial\alpha}$ also tends to zero at $\Delta\alpha \to 0$, which means that we have first-order spatial focusing at the point of intersection of two adjacent trajectories.

Now define two successive increments $\Delta\alpha_1$ and $\Delta\alpha_2$ relative to the principal trajectory. In doing so, we obtain two trajectory equations:

$$\begin{aligned} r_1 &= \psi(z, P, z_i, \alpha) + \frac{1}{1!}\Delta\alpha_1\frac{\partial\psi}{\partial\alpha} + \frac{1}{2!}\frac{\partial\psi}{\partial\alpha}(\Delta\alpha_1)^2 + \frac{1}{3!}(\Delta\alpha)^3\frac{\partial^3\psi}{\partial\alpha^3} + \ldots, \\ r_2 &= \psi(z, P, z_i, \alpha) + \frac{1}{1!}\Delta\alpha_2\frac{\partial\psi}{\partial\alpha} + \frac{1}{2!}\frac{\partial\psi}{\partial\alpha}(\Delta\alpha_2)^2 + \frac{1}{3!}(\Delta\alpha_3)^3\frac{\partial^3\psi}{\partial\alpha^3} + \ldots \end{aligned} \quad (85)$$

Let the trajectories r_1, r_2 intersect with the principal trajectory at the points z_1 and z_2, respectively. Then

$$0 = \frac{1}{1!}(\Delta\alpha_1)\frac{\partial\psi}{\partial\alpha}\bigg|_{z_1} + \frac{1}{2!}(\Delta\alpha_1)^2\frac{\partial^2\psi}{\partial\alpha}\bigg|_{z_1} + \ldots, \quad (86)$$

$$0 = \frac{1}{1!}(\Delta\alpha_2)\frac{\partial\psi}{\partial\alpha}\bigg|_{z_2} + \frac{1}{2!}(\Delta\alpha_2)^2\frac{\partial^2\psi}{\partial\alpha}\bigg|_{z_2} + \ldots \quad (87)$$

Suppose that the values $\Delta\alpha_1$ and $\Delta\alpha_2$ differ by the value ∇; i.e., $\Delta\alpha_2 = \Delta\alpha_1 + \nabla$. Then z_1 and z_2 also differ by a small value Δz; i.e., $z_2 = z_1 + \Delta z$. Now we can write

$$\begin{aligned} \frac{\partial\psi}{\partial\alpha}\bigg|_{z_2} &= \frac{\partial\psi}{\partial\alpha}\bigg|_{z_1} + \Delta z\frac{\partial^2\psi}{\partial\alpha^2}\bigg|_{z_1} + \ldots, \\ \frac{\partial^2\psi}{\partial\alpha^2}\bigg|_{z_2} &= \frac{\partial^2\psi}{\partial\alpha^2}\bigg|_{z_1} + \frac{\partial}{\partial z}\frac{\partial^2\psi}{\partial\alpha^2}\bigg|_{z_1}\Delta z + \ldots. \end{aligned} \quad (88)$$

Taking Eq. (86) into consideration and neglecting the higher orders of Δz, $\Delta\alpha$, and ∇, we obtain from Eq. (87):

$$-\frac{\partial^2\psi}{\partial\alpha^2} = 2!\frac{\Delta z}{V}\frac{\partial}{\partial z}\left(\frac{\partial\psi}{\partial z}\right) + \frac{1}{V}(\Delta\alpha_1 + V)\Delta\frac{\partial}{\partial z}\left(\frac{\partial^2\psi}{\partial\alpha^2}\right) +$$
$$+\frac{1}{V}(\Delta\alpha_1 + V)\Delta z\frac{\partial}{\partial z}\left(\frac{\partial^2\psi}{\partial\alpha^2}\right) + \frac{2!}{3!}2\Delta\alpha_1\frac{\partial^3\psi}{\partial\alpha^3} + \frac{2\cdot 2!}{3!}\Delta\alpha_1\Delta z\frac{\partial}{\partial z}\left(\frac{\partial^3\psi}{\partial\alpha^3}\right) +$$
$$+\frac{2!}{3!}V\frac{\partial^3\psi}{\partial\alpha^3} + \frac{2!}{3!}V\Delta z\frac{\partial}{\partial z}\left(\frac{\partial^3\psi}{\partial\alpha^3}\right) + \frac{2!\Delta z}{3!\,V}(\Delta\alpha_1)^2\frac{\partial}{\partial z}\left(\frac{\partial^3\psi}{\partial z^3}\right) + \ldots \tag{89}$$

It can be easily seen that the second derivative is mainly defined by the value $\frac{\Delta z}{V}$. If this value is zero, the point of trajectory intersection does not depend on α, and $\frac{\partial^2\psi}{\partial\alpha^2}$ represents a first-order value relative to $\Delta\alpha$ and V, or

$$-\frac{\partial^2\psi}{\partial\alpha} = \frac{2!}{3!}(2\Delta\alpha_1 + V)\frac{\partial^3\psi}{\partial\alpha^3} + \ldots \tag{90}$$

Under the condition

$$2\Delta\alpha_1 + V = 0, \tag{91}$$

which implies that $V = -2\Delta\alpha_1$ or $\Delta\alpha_2 = -\Delta\alpha_1$, this second derivative turns to a second-order value with respect to $\Delta\alpha_1$. This defines the technique for searching the mode of second-order focusing: we set the increments $\Delta\alpha_2$ relative to the principal trajectory as having equal absolute values but opposite signs, and use the condition that the points of intersection of those trajectories coincide with the principal trajectory.

By using the technique developed, the lines of first-order spatial focusing in the field-free region at the output of the complete hyperboloidal electrode system have been calculated (Figure 40). It can be deduced that the focusing condition can be realized almost anywhere in the drift space. With z decreasing, the focusing conditions in the drift space will correspond to the condition of first-order focusing at the output end-face hyperboloidal electrode of the axisymmetric electrode system, shown by the dashed line in Figure 40. Now, if we intersect the line of spatial focusing by a plane, we have $\partial r_s/\partial\alpha_{inp} = 0$ at the point of intersection. By placing a narrow annular selection slit at an appropriate spot on this plane, we can ensure that the output of particles have a certain energy.

An ion-optical scheme of the hyperboloidal energy analyzer with selection slits located in the drift space is shown in Figure 41. It includes the axisymmetric hyperboloidal electrode system with the input and output

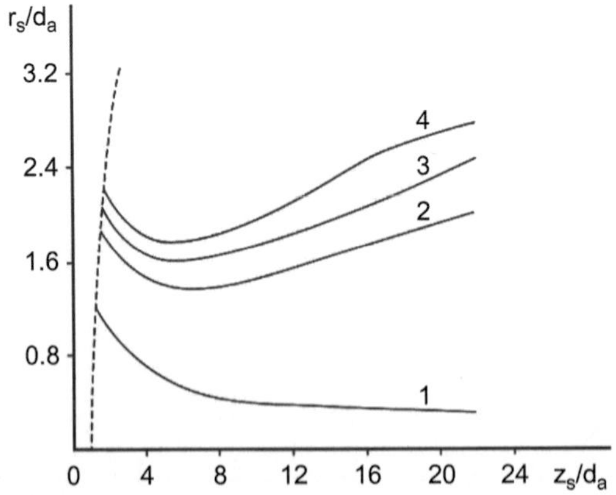

Figure 40 The lines of first-order spatial focusing in the drift space for $z_i = -3d_a$; $\gamma = \sqrt{2}$. Parameters on the curves are as follows: $1 - P' = 1{,}8$; $2 - P' = 2{,}2$; $3 - P' = 2{,}3$; $4 - P' = 2{,}4$ (the dashed line shows the surface of the hyperboloidal output end-face electrode).

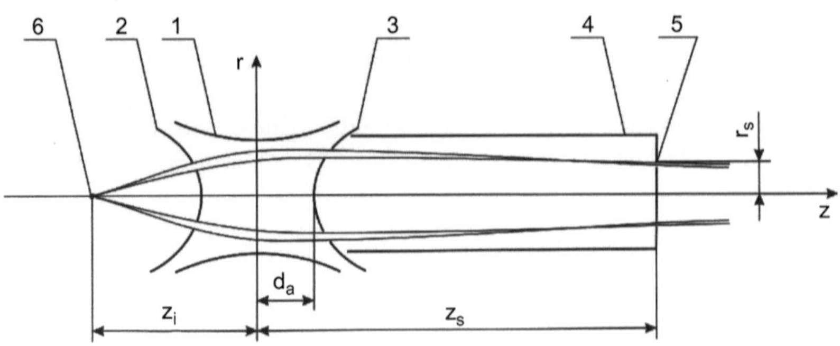

Figure 41 Ion-optical scheme of the energy analyzer based on a hyperboloidal electrode system with selection slits in the drift space.

end-face electrodes (labeled 2 and 3, respectively, in the figure), the annular (controlling) electrode (1), and the electrode (4) with selection slits (5), which is placed in the field-free space. The particles emitted by a point source (6) enter and leave the field region through the circular holes in the end-face electrodes. The locations of the particle source and selection slit relative to the electrode system's center are defined by the values z_i and z_s, respectively.

In the case of the system with selection slits located in the drift space, the calculation of main parameters, including the luminosity and energy resolution, is similar to that described previously. It is of interest to trace the variation of energy analysis parameters along the curve of first-order spatial focusing. The changes in the energy resolution $R_{0.5}$, luminosity $\mathbf{\Omega}$, and input angle $2\Delta\alpha$, obtained for a fixed value of the selection slit width at $P'=2.4$, $z_i=-3d_a$ and $S=0.02d_a$, are shown in Figure 42. With increasing of the input angle α_{inp}, which corresponds to the motion along the focusing curve toward the lesser z values (Figure 40), the energy analyzer resolution decreases, while the luminosity increases. In general, the product of resolution by luminosity remains constant to an accuracy of 15%. Herewith, the analyzer's focal length varies widely, as shown in Figure 42b, in which z_s is the position of the selection slit electrode on the same curve of first-order focusing.

Thus, we can see that placing the selection slits in the drift space of the energy analyzer based on a hyperboloidal electrode system ensures a wide variation of the particle flow geometry with maintaining a high-quality energy analysis.

Comparative analysis of different designs of hyperboloidal axisymmetric energy analyzers. Here, we perform a comparative analysis of different designs of hyperboloidal axisymmetric energy analyzers using the dependencies of the energy resolution on the luminosity (Gurov, Trubitsyun, & Kuksa, 2015). With this aim in view, we select the energy analyzers that possess maximal resolution and luminosity within their type.

As shown by the data presented in Figure 43, maximal values of resolution and luminosity are inherent to the hyperboloidal energy analyzers with

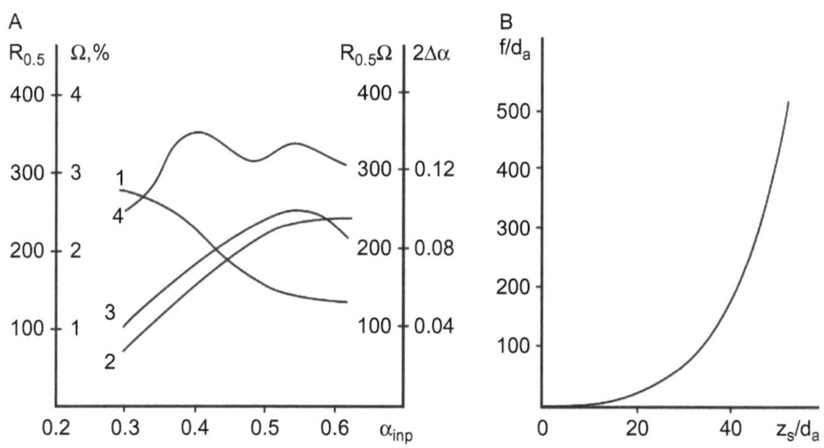

Figure 42 The dependence of the energy analyzer parameters (A) on the particle input angle (A) and the dependence of the energy analyzer's focal length on the selection slit position (B) along the curve of first-order focusing for $P=1.2$(1—resolution $R_{0.5}$; 2—luminosity Ω; 3—opening of input angles $2\Delta\alpha$; 4—product $R_{0.5}\ \Omega$).

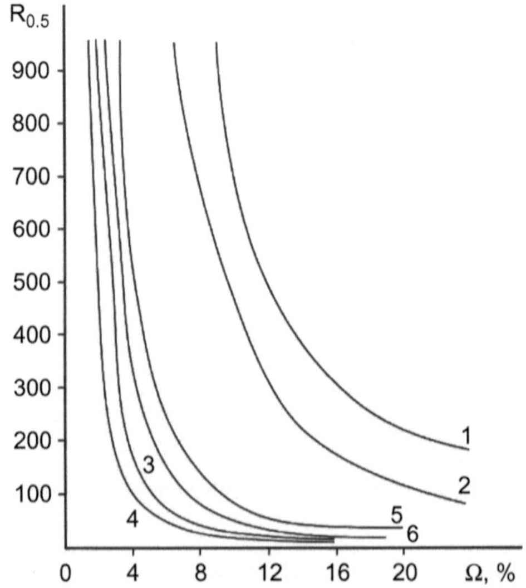

Figure 43 Comparison of various types of axisymmetric hyperboloidal energy analyzers (1, 2—complete electrode system with particle reflection at $\gamma = 0.2$ and $\gamma = \sqrt{2}$, respectively; 3—electrode system with two conical end-face electrodes; 4—complete electrode system with particle focusing; 5, 6—electrode system with a single conical electrode).

particle reflection (curves 1 and 2), which ensure luminosities ranging from 25%–40% at a resolution of 100 for different electrode system geometries.

Next, the type of hyperboloidal axisymmetric energy analyzers that offers superior performance is represented by the energy analyzer with a single conical end-face electrode, providing luminosities ranging from 7.5%–10% at a resolution of 100 for different geometric factor values.

The energy analyzer with two conical end-face electrodes possesses better parameters than the one based on the complete hyperboloidal electrode system. It provides a resolution of 100 with luminosity of 5.5%. The parameters of the energy analyzer with two hyperbolic annular electrodes are close to the parameters of the energy analyzer with two angled electrodes. The energy analyzer based on the complete hyperboloidal electrode system ensures a resolution of 100 at a luminosity of 4%.

The design selection and optimization of the parameters of a real energy analyzer with regard to external structural factors can be performed using the efficiency parameter $Q = R_{0.5}\Omega^2$, which, as we have seen in this chapter, is versatile and can be used with all types of the hyperboloidal axisymmetric energy analyzers considered here. The relevant values of the efficiency parameter can be distributed in descending order as follows:

- Energy analyzer based on a complete hyperboloidal electrode system with particle reflection; $\gamma = 0.2$, $Q = 1.15 \times 10^5$
- Energy analyzer based on a complete hyperboloidal electrode system with particle reflection; $\gamma = \sqrt{2}$; $Q = 4.50 \times 10^4$
- Energy analyzer with a single end-face electrode; $\gamma = 0.2$; $Q = 9.03 \times 10^3$
- Energy analyzer with a single end-face electrode; $\gamma = 1.0$; $Q = 5.63 \times 10^3$
- Energy analyzer with two conical end-face electrodes; $Q = 2.70 \times 10^3$
- Energy analyzer based on a complete hyperboloidal electrode system with particle focusing; $\gamma = 2$; $Q = 1.6 \times 10^3$

A view of the instrumental function allowing estimation of the output signal shape and the possibility of separation of two close energy peaks for the energy analyzer based on a complete hyperboloidal electrode system with particle focusing is shown in Figure 44. It can be seen that the instrumental function's width (and, consequently, the resolution) is varied up to the level of 10^{-2}.

Herewith, the resolution by the level of 10^{-2} is less than the resolution by the level of 0.5 at 1.76 times. The hyperboloidal energy analyzers of other

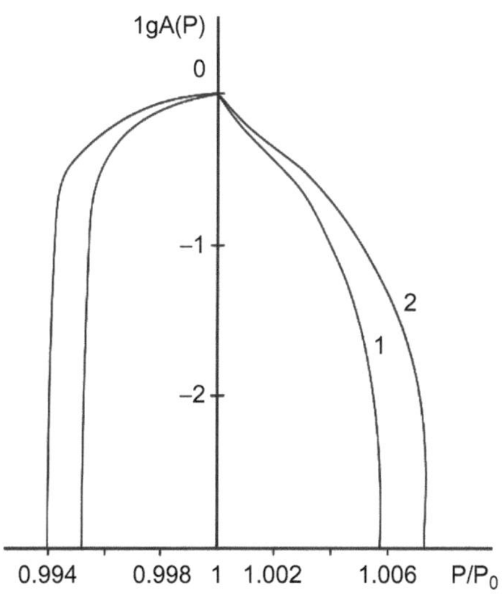

Figure 44 Instrumental function of the energy analyzer based on a complete hyperboloidal electrode system with focusing of particles: $1 - \gamma = \sqrt{2}, z_i = -d_a, P = 1.2$; $2 - \gamma = \sqrt{2}, z_i = -4d_a, P = 1.2$.

types have similarly shaped instrumental functions, with asymmetry with respect to the principal energy.

Experimental samples of hyperboloidal axisymmetric energy analyzers. To test the basic statements of this section, several experimental prototypes of different energy analyzers with hyperboloidal axisymmetric geometry have been designed and an experimental survey performed on them.

The hyperboloidal axisymmetric energy analyzer based on a complete electrode system, which implements the mode of particle focusing, is designed on the basis of the electrode system with the characteristic dimensions $r_a = 19$ mm, $d_a = 19$ mm, and $\gamma = 1.0$. This energy analyzer has the following design parameters:

- Energy resolution at the level of 0.5—60
- Luminosity—4%
- Selection slit width—0.48 mm
- Maximal size of the selection slit—12.59 mm
- Energy coefficient P—1.2
- Distance from the point source to the system's center—8 mm
- Average angle of particle input α_{av}—0.409 rad
- Semi-angle of particle input $\Delta\alpha$—0.04 rad

To test the energy analyzer parameters, a common technique was implemented (Kozlov, 1978). With this aim, a calibrated source of electrons, comprising an 0.4-mm oxide cathode equal in size to that of the emitting surface and two hemispherical stacks for accelerating electrons and ensuring their isotropic distribution in space, was developed and tested. The emission current magnitude can be adjusted from 0 to 100 mA, and the energy spread of electrons is less than 0.3 eV. The accelerating voltage can vary over a wide range, which does not affect the output electron current and its spatial distribution.

Schematic of the experiment included the energy analyzer under study, a calibrated source of electrons, a collector with accelerating and suppressing grids, a measurer of small currents, a recording potentiometer, and a block of power supply. The sweep of recording potentiometer was triggered synchronously with the sweep of the voltage U_{en} on the gate electrode of the energy analyzer.

The behavior of the output signal from the collector for several values of the accelerating voltage U_{acc} is shown in Figure 45. It can be seen that the increase in U_{acc} leads to the peak broadening and shifting toward higher values of U_{en}. Herewith, the peak amplitude varies slightly, which indicates good operation of the calibrated electron source. These energy peaks,

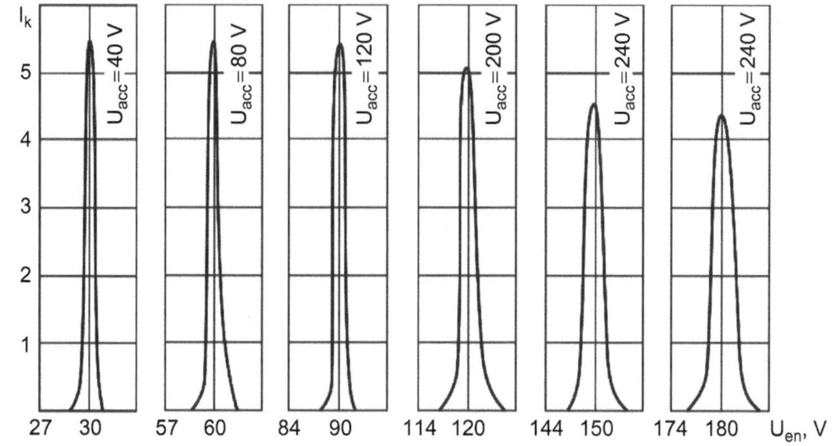

Figure 45 The dependence of the collector current on the control electrode's voltage U_{en} in the energy analyzer for different values of the input energy U_{acc} of charged particles.

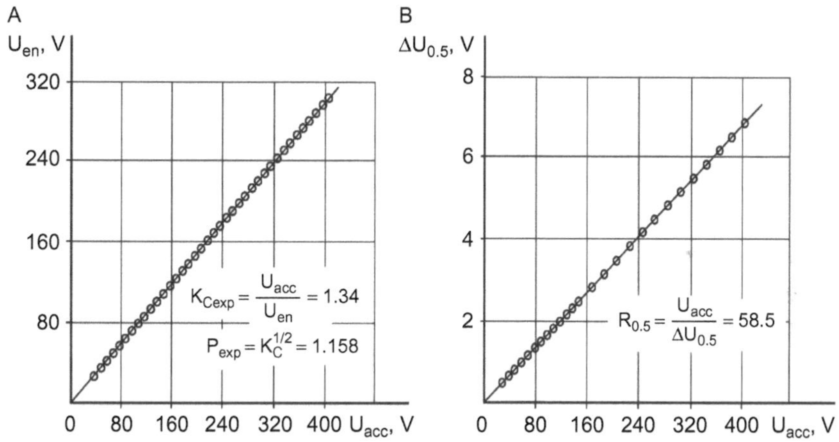

Figure 46 Determination of the coupling coefficient k_c (A) and energy resolution (B).

recorded in a wide range of U_{acc}, serve as a basis for constructing the dependencies of the energy analyzer voltages corresponding to the energy distribution maximum $U_{maxen} = f(U_{acc})$, and the dependencies of the energy distribution half-width at the half-maximum $\Delta U_{0.5} = f(U_{acc})$ on the accelerating voltage. These dependencies are shown in Figure 46 and allow the determination of the coupling factor and the energy analyzer resolution in accordance with experimental data. The experimentally obtained values of $P = 1.158$ and resolution of 58.5 agree quite well with the calculated values.

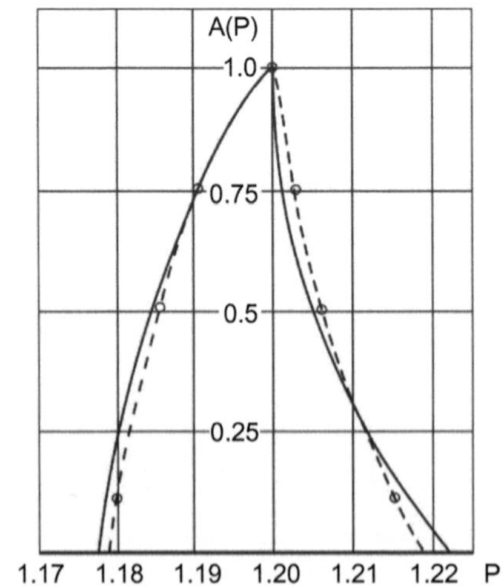

Figure 47 Instrumental function of the energy analyzer (solid line—theoretical dependence, dashed line—experimental dependence).

Experimental verification of the energy analyzer transmittance by sequential measurement of the input and output currents has demonstrated good agreement with the calculated value of 4%.

An important characteristic is the shape of the energy analyzer's instrumental function, which can be obtained experimentally by a monoenergetic beam of charged particles. Figure 47 displays a comparison between a theoretical instrumental function and that obtained experimentally using a calibrated source of particles. A good qualitative agreement between these functions may serve as a reliability criterion for the basic theoretical assumptions underlying the calculations of axisymmetric hyperboloidal energy analyzers.

The experimental prototype of a hyperboloidal axisymmetric energy analyzer with two conical end-face electrodes, in which the mode of particle focusing is implemented, possesses the following design parameters:

- Energy resolution—10
- Luminosity—5%
- Distance from the point source to the system center—47.5 mm
- Coupling coefficient—1.40
- Average angle α_{omH} of particle input—0.355 rad
- Semi-input angle $\Delta\alpha$—0.07

The annular (controlling) electrode of the energy analyzer, representing a hyperboloid of revolution, has the characteristic size $r_a = 19$ mm.

Experimental verification of the energy analyzer parameters differs from that described previously. As a calibration source, a source of electrons with a filament-type, V-shaped, tungsten-rhenium cathode with a diameter of 0.05 mm, which is placed into a modulator with an output aperture diameter of 0.4 mm, is used. Two spherical grids provide acceleration of electrons and their uniform angular distribution. The energy spread of electrons leaving the source does not exceed 1.5 eV, which makes the source suitable for energy analyzer calibration.

Figure 48 shows the dependence of the output signal from the collector for several values of the accelerating voltage U_{acc}. It can be seen that the increase in U_{acc} broadens the energy peak and shifts the peak toward higher values of U_{en}. The output signal amplitude changes slightly. The results of experimental data processing (namely, the dependencies of the energy peak width at the half-maximum $\Delta U_{0.5}$ on the magnitude of accelerating voltage and the magnitude of U_{en} corresponding to the maximum of the energy distribution curve for various U_{acc}, which allows experimental determination of the resolution and luminosity) are shown in Figure 49. The experimentally obtained resolution is $R_{0.5} = 10$, and the coupling coefficient defined from the dependence $U_{en} = f(U_{acc})$ is equal to 1.40, which corresponds to the calculated values.

Evaluation of the system's luminosity is performed by measuring the electron flow injected into the working volume of the energy analyzer and then output therefrom, and it shows fairly good agreement with the calculated value of 5%. Figure 50 shows the calculated and experimental

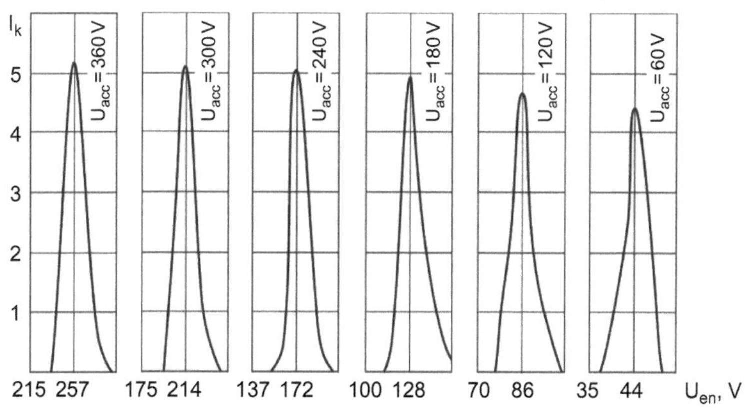

Figure 48 The dependence of the collector current I_k on the control electrode's voltage U_{en} for different input energies of electrons U_{acc}.

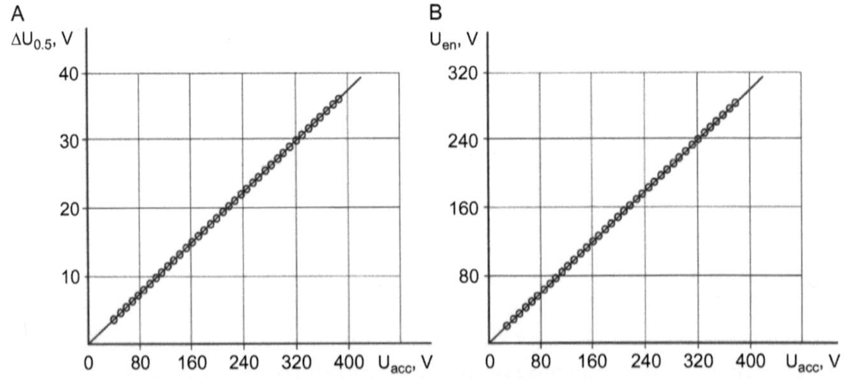

Figure 49 Experimental determination of the resolution (A) and the coupling coefficient (B) in the energy analyzer prototype with two conical end-face electrodes.

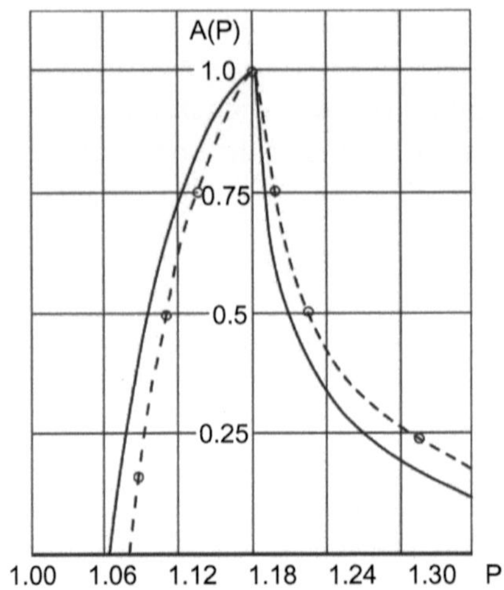

Figure 50 Instrumental function of the energy analyzer with two conical end-face electrodes (solid line—theoretical dependence, dashed line—experimental dependence).

instrumental functions of the energy analyzer under study. A good agreement between the calculation and experiment can be observed.

The energy analyzer based on an axisymmetric hyperboloidal electrode system with selection slits in the drift space has the following design characteristics:

- Energy resolution at the 0.5 level (provided by interchangeable diaphragms with different selection slits)—20 and 50
- Luminosity (for $R_{0.5}=50$)—0.22%

- Coupling coefficient—6.125
- Distance from the point source—77.8 mm
- Distance from the electrode with selection slit—171.6 mm
- Focal length—1750 mm
- Beam diameter at the particle receiver—19 mm

The energy analyzer is constructed on the basis of the hyperboloidal axisymmetric electrode system with $r_a = 11$ mm, $d_a = 7.8$ mm, and geometric factor $= \sqrt{2}$, and is intended to operate with a time-of-flight (TOF) mass spectrometer. Therefore, the experimental testing of the energy analyzer characteristics has been carried out using a TOF mass spectrometer with a laser-plasma ion source (Gurov, 1982). To normalize the ion flow, a system of two coaxial truncated cones with generatrices coinciding with the input angles of particles is placed at the entrance to the energy analyzer.

For experimental determination of the energy analyzer resolution, a well-known technique of energy peak broadening with the increase in the particle input energy has been implemented (Kozlov, 1978). The ion input energy U_{acc} is varied by changing the sample's potential. The input and output end-face electrodes (2, 3) (see Figure 41) of the energy analyzer are grounded. The annular (controlling) electrode (1) is supplied with the linearly varying voltage U_{en}. The energy spectrum is registered using a self-recording potentiometer, the sweep of which is synchronized with the voltage sweep on the controlling electrode. The energy peaks of ions for different accelerating voltages are shown in Figure 51. It is seen that the input energy increase leads to the peak broadening, shift of the peak toward the higher values of U_{en}, and increase in its amplitude. The latter

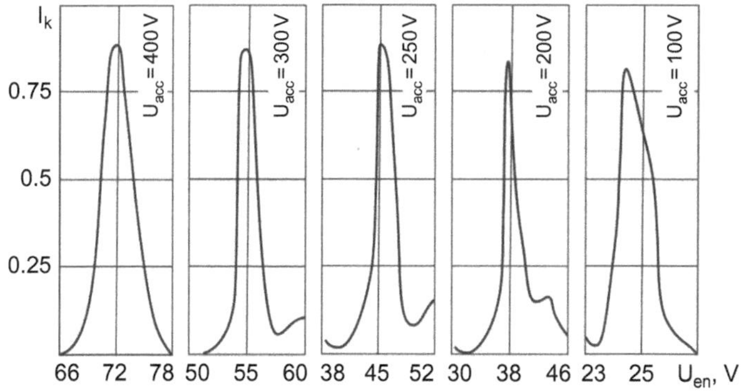

Figure 51 Energy distribution of nickel ions formed under the laser radiation action with flux density $q = 2\ 10^9\ W/cm^2$ for different accelerating voltages U_{acc}.

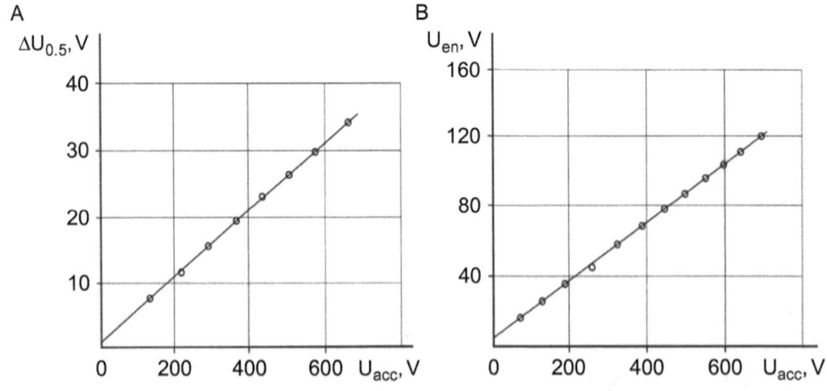

Figure 52 Experimental determination of the energy resolution (A) and coupling coefficient (B) for a hyperboloidal energy analyzer with selection slits located in the drift region.

is connected with the ion source peculiarities. The resulting data are given in Figure 52, where $\Delta U_{0.5}$ represents the energy peak's width at the half-maximum, U_{en} is the energy corresponding to the energy distribution maximum. The tangents of the relevant angles in the dependencies $U_{0.5} = f(U_{acc})$ and $U_{en} = f(U_{acc})$ give the experimental values of the energy resolution and coupling coefficient, respectively. The experimentally determined values of the energy resolution $R_{0.5} = 18.5$ and the coupling coefficient $k_c = 6.125$ are well correlated with the calculated data.

To test the coupling coefficient value, a different TOF technique has been elaborated, the essence of which consists in measuring the TOF $t_{\text{п}r}$ of the ions of certain mass through the drift tube of the length L (Gurov, 1982). Herewith, the coupling coefficient k_c takes the form

$$k_c = \frac{L^2}{2\left(\frac{e}{m_i}\right)\left(\frac{n_i}{A}\right) U_{en} t_f^2}, \tag{92}$$

where e is the electron charge, m_i is the mass of atomic unit, A is the atomic mass of ion, n_i is the ionization degree, and U_{en} is the voltage on the annular (controlling) electrode of the energy analyzer.

Nickel is used as the target material. The voltage U_{en} is varied in the range that allows observation of the mass spectrum of singly and doubly charged ions of nickel. The TOF is recorded by a wideband oscilloscope, the sweep of which is synchronized with the laser pulse action. The results of the determination of the coupling coefficient are presented in Figure 53, in which the

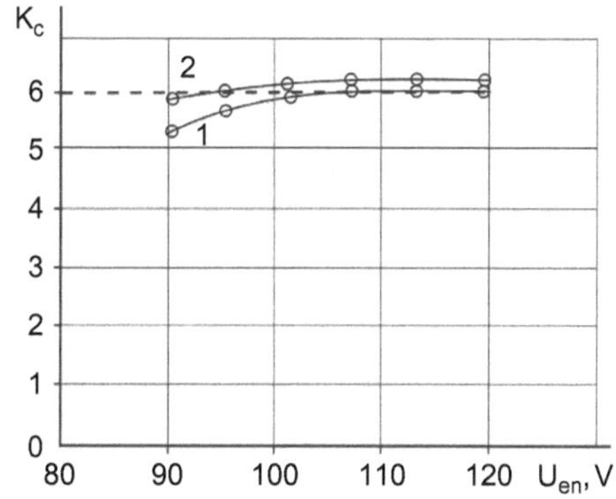

Figure 53 Experimental determination of the coupling coefficient k_c using the TOF technique developed: $U_{acc}=600\ V$, $1-Ni^+$; $2-Ni^{++}$ (the dashed line stands for the theory; ° ° ° ° stands for the experiment).

upper curve corresponds to the calculation of k_c according to the TOF of doubly charged ions and the lower curve reflects the case of singly charged ions of nickel. The dashed line indicates the theoretical value of k_c. The experimentally found average value $k_c=6.22$ is in good agreement with the calculated value of 6.125.

The estimated collection efficiency of the energy analyzer is about $10^{-4}/10^{-5}$, which agrees qualitatively with the calculated value, given the specificity of the laser-plasma source. It should be emphasized that this value of collection efficiency, obtained experimentally, is rather high for laser mass spectrometry (Sheretov & Gurov, 1982). The qualitative agreement between the experimental and calculated values of collection efficiency may serve as an indirect confirmation of the correctness of the focal length as applied to the system developed.

External views of the energy analyzer designs based on the hyperboloidal electrode systems of different types and their design features are shown in Figures 54–56.

The working volume of the hyperboloidal axisymmetric energy analyzer with particle focusing (Figure 55) consists of three field-generating electrodes (labeled 1, 2, and 3 in the figure) having a hyperbolic profile (Sheretov & Gurov, 1980). The energy analyzer with conical end-face electrodes comprises the hyperbolic annular electrode (1) and two conical electrodes coinciding with the system's asymptotes. All these electrodes are manufactured from copper using electrochemical deposition. The relative position of the electrodes and their electrical insulation are determined by the ceramic insulators (5). The tolerances for profile manufacture and setting

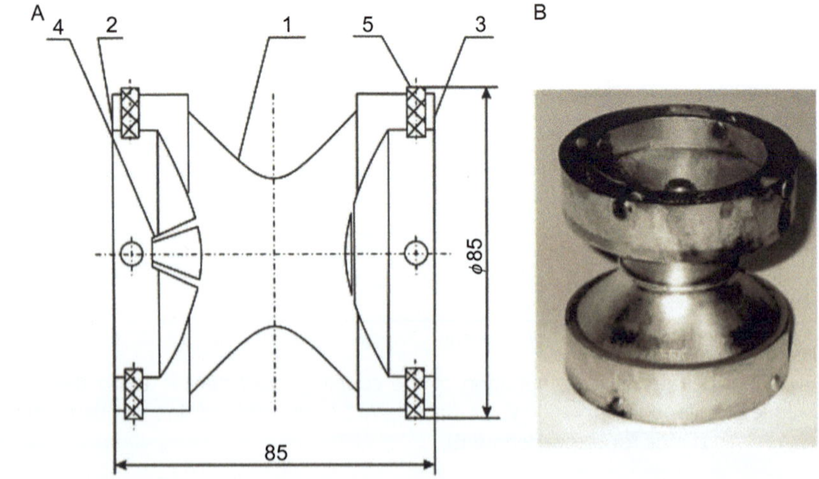

Figure 54 Structure of electrodes (A) and external view (B) of the energy analyzer based on a complete hyperboloidal electrode system with $\gamma = 1{,}0$; $r_a = d_a = 19$ mm. 1, 2, 3 - field-generating electrodes, 4 – truncated cones, 5 – ceramic insulators. (See the color plate.)

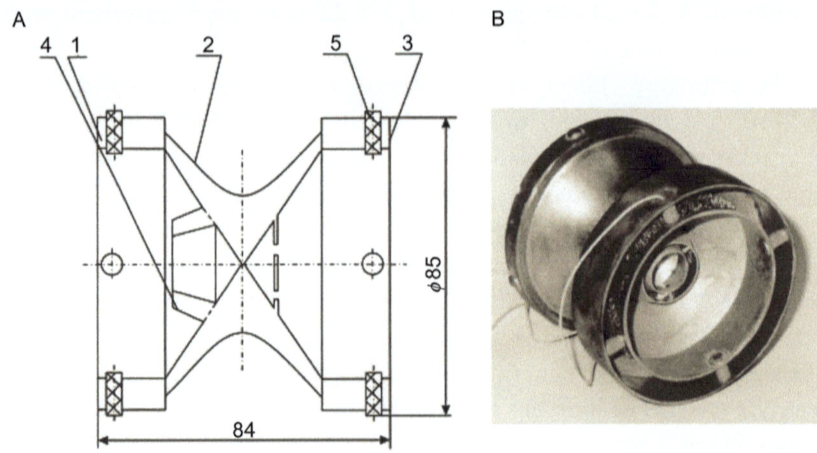

Figure 55 Structure of electrodes (A) and external view (B) of the energy analyzer based on an axisymmetric hyperboloidal electrode system with two conical end-face electrodes ($\gamma = 1{,}0$; $r_a = d_a = 19$ mm). 1, 2, 3 - field-generating electrodes, 4 – truncated cones, 5 – ceramic insulators. (See the color plate.)

the relative position of electrodes do not exceed 10μ. As the material for insulators, tubular ceramics with a 4.5-mm diameter are used.

The end-face electrodes (2, 3) have annular slits in the form of four sectors for the input of charged particles into the working volume. A system of input truncated cones (4) serves to normalize the input angles

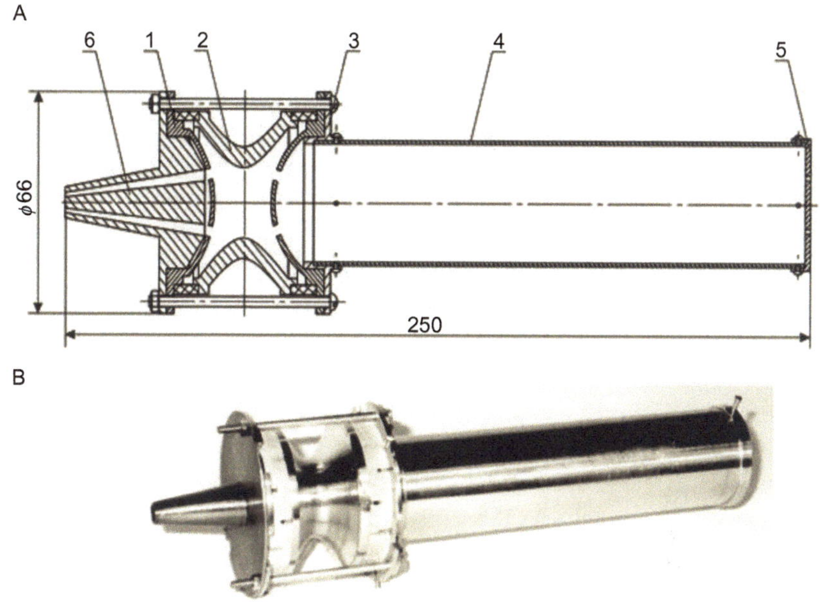

Figure 56 Energy analyzer based on the hyperboloidal electrode system with selection slits located in the drift space ($\gamma = \sqrt{2}$; $r_a = 11$ mm, $d_a = 7.78$ mm). 1, 2, 3 – electrodes of hyperboloidal electrode system, 4 – drift tube, 5 – selection slits. (See the color plate.)

of the particles emitted from a point source (Sheretov et al., 1985). Normalization of input angles is provided by the fact that the generatrices of the internal and external cones coincide with the maximum and minimum input angles. The vertex of truncated cones is located at the point corresponding to the particle source. The degree of truncation is selected from structural considerations. To minimize the field distortion, the input annular slit is covered by a fine-structure mesh with high transparency. The field distortion caused by the presence of selection slit can be neglected since the slit sizes do not exceed 0.3–0.5 mm. Overall, the dimensions of the energy analyzers described do not exceed 88 × 80 mm; those analyzers are made on the basis of the axisymmetric hyperboloidal electrode system with $\gamma = 1$ and characteristic dimensions $r_a = d_a = 19$ mm.

The hyperboloidal axisymmetric energy analyzer with selection slits in the drift space (Figure 56) is performed on the hyperboloidal electrode system with $\gamma = \sqrt{2}$, $r_a = 11$ mm, $d_a = 7.78$ mm. Structurally, it consists of two components—the hyperboloidal electrode system (1–3) and the drift tube (4) with selection slits (5) located on the output aperture. The fixing, insulation, and relative positioning of the field-generating electrodes are provided by calibrated annular ceramics. The drift tube (4) is fastened via a connector to the output end-face electrode (3), which ensures the tube coaxiality with the electrode system and the accuracy of positioning the selection slits.

All the system components are manufactured from stainless steel. The manufacture accuracy of the field-generating electrodes and assembling accuracy of the entire system are not greater than 0.05 mm. To minimize field distortion in the energy analyzer, the annular sector slits for input and output of charged particles, which are manufactured using the electrical discharge method, are covered by a high-transparency, fine-structure mesh. To normalize the conditions of inputting the particles into the beam-forming system, the abovementioned structure consisting of two truncated cones was used. The beam-forming system design allows convenient and fast assembly. The overall dimensions of the beam-forming system for the laser-plasma ion source and the TOF mass spectrometer, with a 1.75-m-long drift tube, are 66×200 mm.

The unquestionable advantage of the created hyperboloidal axisymmetric energy analyzers is their high analytical characteristics, easy access to the input and output apertures, the opportunity to widely change the input and output particle parameters, and their small size and weight.

The following general conclusions can be derived from this chapter:

- Static electric fields produced by hyperboloidal electrode systems possess considerable energy dispersion. Depending on the initial conditions of the particle input, the modes of particle return (reflection) and focusing (transit) can be implemented in those fields.
- Under certain initial conditions of inputting the charged particles into the field, there is an opportunity to implement the modes of first- and second-order spatial focusing in the plane of the output electrode in a hyperboloidal electrode system. Herewith, implementation of second-order focusing is possible only in the mode of particle return (reflection). In the transit mode, only first-order focusing is possible.
- Energy analyzers of charged particles can be created using all types of axisymmetric hyperboloidal electrode systems with implementation of the modes of return and reflection and arbitrary positioning of the ion source and selection slit.
- For comparison of different designs of hyperboloidal energy analyzers, optimization of their characteristics, and carrying out engineering calculations of real prototypes, it is advisable to use the efficiency parameter, defined as a product of energy resolution by squared luminosity. This

parameter represents a versatile characteristic of all types of the hyperboloidal energy analyzers discussed in this chapter.

- Numerical modeling has shown that the highest luminosity among the hyperboloidal energy analyzers is inherent in the energy analyzers based on the complete electrode system with particle return and the energy analyzers with conical end-face electrodes. By their parameters, these energy analyzers considerably exceed the known types of energy analyzers.
- The output flows of charged particles in a hyperboloidal axisymmetric electrode system may have angles of convergence both of positive and negative signs, which are completely determined by the system geometry and initial conditions of inputting the particles into the field. However, in the regions of the output electrode with the condition of first-order spatial focusing fulfilled, only negative velocities of output particles are realized. The focal length of such a forming system does not exceed several characteristic dimensions of the electrode system.
- To increase the focal length of the energy analyzer based on a single hyperboloidal electrode system, the selection slit should be placed within the field-free region. The position of selection slits in the drift space is defined by the lines of first-order spatial focusing, which can be found using the calculation technique proposed. When moving along the curve of spatial focusing from the output end-face electrode, the focal length of the forming system increases, while the energy analysis parameters (resolution and luminosity) vary insignificantly.
- Experimental testing of the energy analyzer based on a hyperboloidal axisymmetric electrode system with selection slits in the drift space has shown a good agreement between theoretical and experimental data and the correctness of the main provisions that form the basis of the calculations performed.

CHAPTER THREE

Approximate-Analytical Method of Calculating the Charged Particle Trajectories in Electrostatic Fields

Contents

A class of multipole-cylindrical axisymmetric Laplace fields based on the synthesis of multipoles and cylindrical-type electrostatic fields was proposed by Zashkvara (1995) and Zashkvara and Tyndyk (1996). Studies on electron-optical properties of this class of synthetized multipole-cylindrical fields are of practical interest for implementation of new energy analyzers of charged particle flows. The main difficulty in the development of such systems is a sufficiently accurate description of charged particle trajectories. A solution to this problem is necessary for estimating the aberrations that define the limiting resolution of the electron-optical systems in question.

In Zashkvara and Tyndyk (1999) and Zashkvara, Ashimbaeva, and Chokin (2002), an approximate-analytical method of calculating the charged particle trajectories was developed to determine the electron-optical parameters of mirror-type energy analyzers based on an axisymmetric multipole-cylindrical field. By using this mathematical model, calculation of charged particle trajectories in a hexapole-cylindrical field and evaluation of electron-optical parameters of relevant energy analyzers were performed by Ashimbaeva, Chokin, Saulebekov, and Assylbekova (2004), Ashimbaeva,

Advances in Imaging and Electron Physics, Volume 192
ISSN 1076-5670
http://dx.doi.org/10.1016/bs.aiep.2015.08.003

Chokin, and Saulebekov (2005, 2008), Ashimbaeva, Chokin, Saulebekov, and Kambarova (2012b).

In this chapter, we propose an approximate-analytical method of superpositioning the truncated power series to describe the charged particle trajectories in an electrostatic mirror-type energy analyzer based on the axisymmetric hexapole-cylindrical field. We also analyze the trajectory calculations in such a field representing a combination of the circular cylindrical mirror and hexapole; herewith, the contribution of the latter can be varied by using a certain coefficient. Introducing an additional free parameter allows enhancement of the basic electron-optical characteristics of the system as a whole.

Let us analyze the superposition structure of a multipole (Zashkvara, 1995; Zashkvara & Tyndyk, 1996) and a cylindrical electrostatic field. In this superposition, the multipole's central circumference coincides with having no equipotential line of the logarithmic field. Consider the superposition of a cylindrical hexapole (Zashkvara & Tyndyk, 1996) and a cylindrical field. The variants shown in Figure 1 correspond to the cases $U(r, z) = \ln r - U_h(r, z)$ (A), $U(r, z) = \ln r + U_h(r, z)$ (B), and $U(r, z) = \frac{5}{2} \ln r - U_h(r, z)$ (C).

A

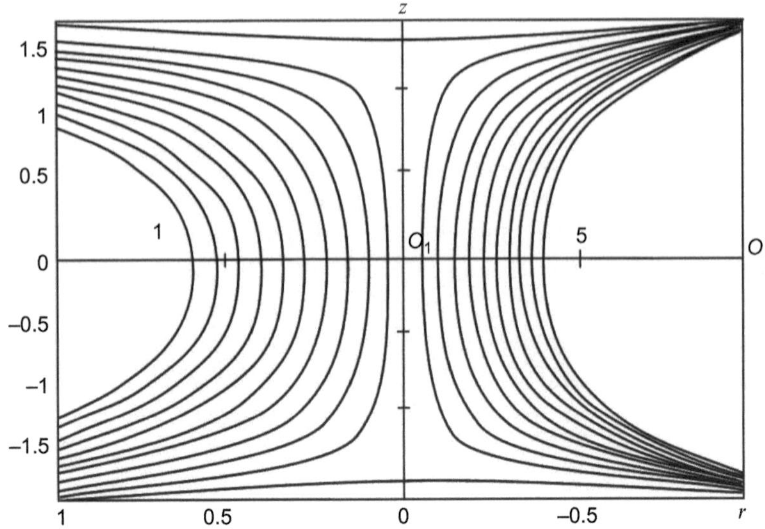

Figure 1 Equipotential lines of the fields representing a superposition of the hexapole and cylindrical field (A) $U(r,z) = \ln(1+r) - U_h(r,z)$.

(Continued)

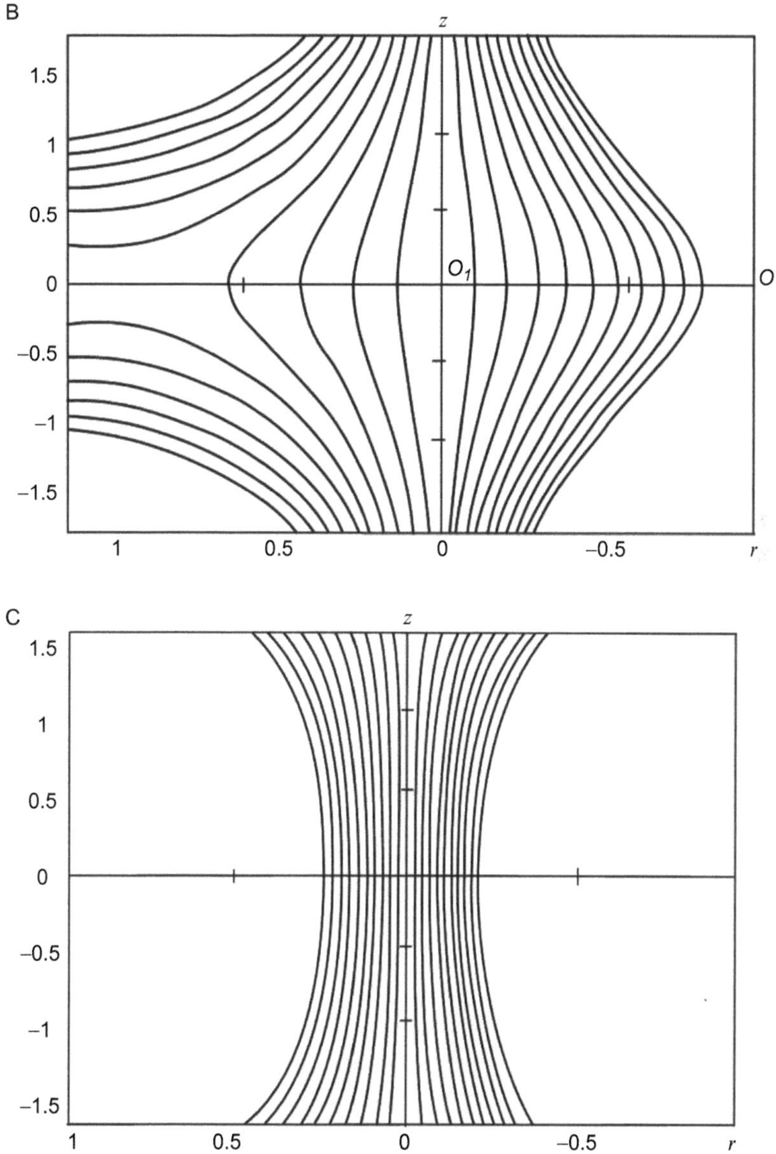

Figure 1—Cont'd (B) $U(r,z) = \ln(1+r) + U_h(r,z)$, (C) $U(r,z) = \frac{5}{2}\ln(1+r) - U_h(r,z)$.

The schemes of axisymmetric electrostatic mirror-type analyzers, in the frame of which these variants of the hexapole and cylindrical field's superposition can be implemented, are discussed next. The energy analyzer comprises an axisymmetric cylindrical electrode and a deflecting electrode that possesses a generatrix coinciding with one of the field's equipotential lines. The field to deflect the charged particles is formed between those electrodes.

The outer electrode may have either a convex (Figure 1b) or a concave (Figure 1c) shape. The case shown in Figure 1a also represents a certain interest owing to the presence of the electrodes that can be used to protect the working area from the fringe effects. Sections 3 to 5 in this chapter are dedicated to the calculation of the hexapole-cylindrical analyzers as applied to the variants of superposition shown in Figure 1.

1. INTEGRO-DIFFERENTIAL EQUATION OF CHARGED PARTICLE TRAJECTORIES IN THE ELECTROSTATIC HEXAPOLE-CYLINDRICAL FIELD $U(r,z) = \ln r + \gamma U_h(r,z)$

The following superposition of the cylindrical field $U_c = \ln r$ and a circular hexapole is of primary interest:

$$U(r, z) = \ln r + \gamma U_h(r, z), \tag{1}$$

where

$$U_h(r, z) = \frac{1}{2}\left\{\left[z^2 - \frac{r^2}{2} - \frac{1}{2}\right] \ln r + \frac{r^2}{2} - \frac{1}{2}\right\} \tag{2}$$

represents the circular hexapole, and γ is the weight component. The motion of charged particles in the hexapole-cylindrical field [Eq. (1)] is schematically shown in Figure 2. The field is formed in the space between two axisymmetric coaxial cylindrical electrodes, in which the inner cylinder has the radius r_0 and is placed at ground potential, while the outer electrode with a curved profile carries the potential U_0, the sign of which coincides with the sign of charge of the particles moving in the field [Eq. (2)].

The problem is solved in cylindrical coordinates, with an angular coordinate not taken into account due to axial symmetry. It is convenient to shift the coordinate system origin to the trajectory's vertex located at point m and pass over to the coordinates x and ξ, as shown in Figure 2.

According to the previously adopted procedure (Zashkvara et al., 2002), all linear dimensions are normalized by the radius r_0 of the inner cylindrical electrode:

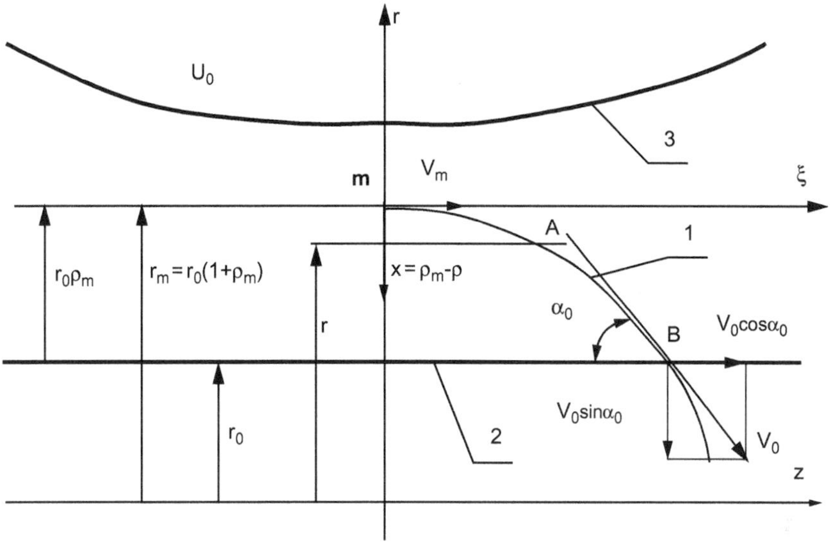

Figure 2 A charged particle trajectory in the electrostatic hexapole-cylindrical field $U(r,z) = \ln r + \gamma U_h(r,z)$.

$$x = \frac{r_m - r}{r_0} = \rho_m - \rho,\ \frac{r}{r_0} = \frac{r_0 + r_0\rho}{r_0} = 1 + \rho,\ \xi = \frac{z}{r_0}. \tag{3}$$

From the theorem on kinetic energy variation, for an arbitrary point A and the vertex point m of the trajectory, we obtain

$$\frac{m}{2}\left(\dot{x}^2 + \dot{z}^2\right) - \frac{m}{2}V_m^2 = q(U_m - U) = q[U(0,0) - U(x,z)]. \tag{4}$$

Taking into account that $\dot{x} = \frac{dx}{dz}\frac{dz}{dt} = x'\dot{z}$, we can rewrite Eq. (4) in the following form:

$$\frac{m}{2}\dot{z}^2\left(x'^2 + 1\right) = \frac{m}{2}V_m^2 + q[U(0,0) - U(x,z)], \tag{5}$$

where

$$x'^2 = \frac{2}{m\dot{z}^2}\left\{\frac{m}{2}V_m^2 - \frac{m}{2}\dot{z}^2 + q[U(0,0) - U(x,z)]\right\} \tag{6}$$

and

$$\left(\frac{d\xi}{dx}\right)^2 = \frac{\frac{m}{2}\dot{z}^2}{\frac{m}{2}V_m^2 - \frac{m}{2}\dot{z}^2 + q[U(0,0) - U(x,z)]}, \tag{7}$$

since

$$\frac{dx}{dz}=\frac{dx}{d\xi}=\frac{1}{d\xi/dx}=\frac{1}{\xi'}. \tag{8}$$

For the longitudinal component of the charged particle motion, we have

$$\frac{m}{2}\dot{z}^2=\frac{m}{2}V_m^2+q\int_0^Z E_Z dz, \tag{9}$$

and, with regard to the relation

$$U(x,\xi)=U_0 g(x,\xi), \tag{10}$$

we obtain

$$E_r=\frac{U_0}{r_0}\varepsilon_1(x,\xi),\ \varepsilon_1=-\frac{\partial g}{\partial \rho} \tag{11}$$

$$\frac{m}{2}\dot{z}^2=\frac{m}{2}V_m^2+qU_0\int_0^\xi \varepsilon_2 d\xi=\frac{m}{2}V_m^2+qU_0\int_0^x \varepsilon_2\frac{d\xi}{dx}dx \tag{12}$$

$$\frac{m}{2}V_m^2-\frac{m}{2}\dot{z}^2=-qU_0\int_0^x \varepsilon_2\xi' dx. \tag{13}$$

Transformation of the denominator in Eq. (7), while taking into account Eqs. (10) and (13), gives

$$\frac{m}{2}V_m^2-\frac{m}{2}\dot{z}^2+qU_0[g(0,0)-g(x,\xi)]=qU_0[g(0,0)-g(x,\xi)]-qU_0\int_0^x \varepsilon_2\xi' dx. \tag{14}$$

Substituting into Eq. (13) $\dot{z}_B=V_0\cos\alpha_0$, $x=\rho_m$, at the point B, we obtain

$$\frac{m}{2}V_m^2=\frac{m}{2}V_0^2\cos^2\alpha_0-qU_0\int_0^{\rho_m}\varepsilon_2\xi' dx. \tag{15}$$

Substitution of Eq. (15) into Eq. (13) brings us to the relation

$$\frac{m}{2}\dot{z}^2=\frac{m}{2}V_0^2\cos^2\alpha_0-qU_0\int\limits_0^{\rho_m}\varepsilon_2\xi' dx+qU_0\int\limits_0^{x}\varepsilon_2\xi' dx,$$

where we obtain

$$\frac{m}{2}\dot{z}^2=\frac{m}{2}V_0^2\cos^2\alpha_0-qU_0\int\limits_x^{\rho_m}\varepsilon_2\xi' dx. \tag{16}$$

Finally, using Eqs. (14) and (16), from Eq. (7), we derive

$$\left(\frac{d\xi}{dx}\right)^2=\frac{\frac{m}{2}V_0{}^2\cos^2\alpha_0-qU_0\int\limits_x^{\rho_m}\varepsilon_2\xi' dx}{qU_0[g(0,0)-g(x,\xi)]-qU_0\int\limits_0^{x}\varepsilon_2\xi' dx},$$

which is equivalent to

$$\left(\frac{d\xi}{dx}\right)^2=\frac{P^2 ctg^2\alpha_0-\int\limits_x^{\rho_m}\varepsilon_2\xi' dx}{g(0,0)-g(x,\xi)-\int\limits_0^{x}\varepsilon_2\xi' dx}, \tag{17}$$

where

$$P=\sqrt{{}^{mV_0^2}\big/_{2qU_0}}\sin\alpha_0 \tag{18}$$

is the mirror's reflection parameter, where m, q, and V_0 are the mass, charge, and velocity of particles, respectively.

The electrostatic field in question represents a superposition of the cylindrical field $g_c = \ln(1+\rho)$ and the circular hexapole field:

$$g_h=\frac{1}{2}\left\{\left[\xi^2-\frac{1}{2}\left(1+\rho^2\right)-\frac{1}{2}\right]\ln(1+\rho)+\frac{1}{2}\left(1+\rho^2\right)-\frac{1}{2}\right\}, \tag{19}$$

$$g(\xi,\rho)=g_c+\gamma g_h$$
$$=\ln(1+\rho)+\frac{\gamma}{2}\left\{\left[\xi^2-\frac{1}{2}(1+\rho^2)-\frac{1}{2}\right]\ln(1+\rho)+\frac{1}{2}(1+\rho^2)-\frac{1}{2}\right\},$$

$$g(\xi,\rho)=\frac{\gamma}{2}\left\{\left[\xi^2-1+\frac{2}{\gamma}-\rho-\frac{1}{2}\rho^2\right]\ln(1+\rho)+\frac{1}{2}\rho^2+\rho\right\}. \qquad (20)$$

Thus, Eq. (17) together with relations (18), (20) represent an integro-differential equation describing the charged particle trajectories in the hexapole-cylindrical field [Eq. (20)].

2. CALCULATION OF CHARGED PARTICLE TRAJECTORIES IN THE ELECTROSTATIC HEXAPOLE-CYLINDRICAL FIELD $U(r,z) = \ln r + \gamma U_h(r,z)$

Consider the integro-differential equation [Eq. (17)] describing the charged particle trajectories in the hexapole-cylindrical field. This equation does not permit the separation of variables and the representation of a trajectory in terms of elementary functions or quadratures. The basic trajectory of a particle in an electrostatic axisymmetric mirror has a vertex in the field region and cannot be generally described by an elementary function. Such a planar trajectory, which Zashkvara et al. (2002) called a "return" trajectory, consists of two branches that are symmetrical relative to the trajectory's vertex. Compared to the direct use of numerical methods, the return trajectory calculation can be greatly facilitated if we apply an approximate-analytical approach based on the use of an optimally selected superposition of the truncated power series.

Let us construct a solution to Eq. (17) in the form

$$\xi(x)=\sqrt{x}\sum_{n=0}^{\infty}C_n x^n. \qquad (21)$$

In doing so, the problem is reduced to finding coefficient C_n in Eq. (21). Further, we have to find sequentially the terms of Eq. (17) in the form of a power series, and after that, equate the terms having equal powers of x to determine the coefficients of the series [Eq. (21)]. This procedure is described in detail in Appendix 1.

Considering the system of Eqs. (A1.25)–(A1.28) and omitting intermediate transformations, we arrive at a recurrent relation for coefficient h_i that depends on the sought-for coefficient C_i:

$$h_{i+1} = \frac{1}{b_1 - e_1/2}\left[-\frac{4}{i+1}b_{i+1} - \sum_{m=0}^{i} h_m\left(-\frac{1}{2}e_{i+2-m} + \frac{1}{i+2-m}b_{i+2-m}\right)\right]. \tag{22}$$

The first part of Eq. (A1.25) can be solved separately, and using Eqs. (A1.26)–(A1.28), we obtain

$$C_0^2 = 4\frac{P^2\text{ctg}^2\alpha_0 + \gamma F(\rho_m)}{\frac{1}{1+\rho_m} - \gamma\left[\frac{1+\rho_m}{2}\ln(1+\rho_m) + \frac{\rho_m^2}{4(1+\rho_m)} - \frac{\rho_m}{2}\right]}. \tag{23}$$

To find the subsequent coefficient C_i, we can use the recurrent relation [Eq. (22)] and Eqs. (A1.26)–(A1.28). In particular, at $i=0$, we obtain

$$h_1 = \frac{1}{b_1 - e_1/2}\left[-4b_1 - h_0\left(-\frac{e_2}{2} + \frac{b_2}{2}\right)\right], \tag{24}$$

Replacing h_0, h_1, e_1, e_2, b_1, and b_2 with the relevant expressions in Eqs. (A1.26)–(A1.28) and resolving them with respect to C_1, we get

$$C_1 = \frac{C_0}{6}\frac{9\ln(1+\rho_m) + 1 + \frac{C_0^2 - 1}{1+\rho_m} - \frac{1}{2(1+\rho_m)^2}\left(\rho_m{}^2 - \frac{4}{\gamma}\right)}{2(1+\rho_m)\ln(1+\rho_m) + \frac{1}{1+\rho_m}\left(\rho_m{}^2 - \frac{4}{\gamma}\right) - 2\rho_m}. \tag{25}$$

At $i=1$, we obtain the corresponding expressions for h_2 and C_2:

$$h_2 = \frac{1}{b_1 - e_1/2}\left[-2b_2 - h_0\left(-\frac{e_3}{2} + \frac{b_3}{3}\right) - h_1\left(-\frac{e_2}{2} + \frac{b_2}{2}\right)\right], \tag{26}$$

$$C_2 = \frac{C_0}{10}\left\{-\frac{11\,C_1}{2\,C_0}\ln(1+\rho_m) - \frac{3\,C_1}{2\,C_0} + \frac{1}{1+\rho_m}\left[1 - \frac{C_0C_1}{3} - \frac{3\,C_1}{2\,C_0}(C_0^2 - 1)\right]\right.$$

$$\left. - \frac{C_0^2 - 1}{6(1+\rho_m)^2} + \frac{1}{4(1+\rho_m)^2}\left[3\frac{C_1}{C_0} + \frac{1}{3(1+\rho_m)}\right](\rho_m{}^2 - 4/\gamma)\right\} / \{\rho_m/2 \tag{27}$$

$$-\frac{1}{2}(1+\rho_m)\ln(1+\rho_m) - \frac{1}{4}\frac{1}{1+\rho_m}(\rho_m{}^2 - 4/\gamma)\} - \frac{9}{10}\frac{C_1^2}{C_0}.$$

The next step is to determine the coefficients h_3 and C_3 at $i=2$:

$$h_3 = \frac{1}{b_1 - e_1/2}\left[-\frac{4}{3}b_3 - h_0\left(-\frac{e_4}{2}+\frac{b_4}{4}\right) - h_1\left(-\frac{e_3}{2}+\frac{b_3}{3}\right) - h_2\left(-\frac{e_2}{2}+\frac{b_2}{2}\right)\right] \tag{28}$$

$$\begin{aligned}
C_3 = &\frac{1}{7C_0}\frac{1}{\rho_m - (1+\rho_m)\ln(1+\rho_m) - \frac{1}{2(1+\rho_m)}(\rho_m{}^2 - 4/\gamma)} \\
&\times\Bigg\{-4\left(C_0C_2 + C_1^2/2\right)\ln(1+\rho_m) + \frac{8}{3}\frac{C_0C_1}{1+\rho_m} + \frac{C_0^2}{3(1+\rho_m)^2} \\
&-C_0^2\Bigg[\left(2C_0C_1 + C_1^2\right)\frac{1}{8(1+\rho_m)} \\
&+\frac{C_0C_1}{4(1+\rho_m)^2} + \frac{1}{24(1+\rho_m)^2} + \frac{C_0^2-1}{8(1+\rho_m)^3} - \frac{1}{16(1+\rho_m)^4}(\rho_m{}^2 - 4/\gamma)\Bigg] \\
&-6C_0C_1\left[\frac{C_0C_1}{3(1+\rho_m)} + \frac{C_0^2-1}{6(1+\rho_m)^2} - \frac{1}{12(1+\rho_m)^3}(\rho_m{}^2 - 4/\gamma)\right] \\
&-\frac{1}{4}\left(10C_0C_2 + 9C_1^2\right) \\
&\times\left[1 - \ln(1+\rho_m) + \frac{C_0^2}{1+\rho_m} - \frac{1}{1+\rho_m} - \frac{1}{2(1+\rho_m)^2}(\rho_m{}^2 - 4/\gamma)\right]\Bigg\} \\
&-\frac{15}{7}\frac{C_1C_2}{C_0}.
\end{aligned} \tag{29}$$

Taking into consideration the complexity of calculating coefficient C_i, we restrict ourselves to the already-calculated coefficients C_0, C_1, C_2, and C_3.

3. ELECTRON-OPTICAL PROPERTIES OF THE HEXAPOLE-CYLINDRICAL ENERGY ANALYZER WITH END-FACE ELECTRODES ($\gamma = -1$)

One of the challenges of creating an actual cylindrical spectrometer is the protection of the device's working area from the fringe field action. In a real spectrometer, the length of cylindrical electrodes in the symmetry axis direction is limited, and the boundary conditions at the end faces can significantly affect the field distribution in the region of charged particle motion.

To maintain a sufficient accuracy of approximation with respect to a cylindrical field in the analyzing part of the spectrometer, it is required either to lengthen the cylindrical electrodes or to take special measures to protect the working area without increasing the longitudinal dimensions of electrodes. From the practice of designing the electrostatic spectrometers, we have found a method of protecting the working area by means of introducing additional electrodes, the potentials of which are distributed in accordance with a certain law. In particular, the fringe fields in the spherical-type sectorial electrostatic spectrometer described by Zigban et al. (1971) are compensated by three curved rods carrying a specially calculated potential to approximate the spherical field distribution in the working area. This method of protection is the most simple and effective. In the well-known electrostatic cylindrical energy analyzer (Zashkvara, Korsunskii, & Kosmachev, 1966), the fringe field compensation is performed using a distributed potential. The characteristic region of fringe field penetration can be estimated as 2–3 distances between the rings. The greater the number of circular shield electrodes, and the smaller their thickness is, the higher is the accuracy of maintaining a correct electrostatic field distribution at the end faces.

To solve the problem of protecting the working area, we have proposed an electrostatic mirror-type energy analyzer based on the hexapole-cylindrical field generated by the circular cylindrical mirror and hexapole. The longitudinal dimensions of the analyzer are limited by the end-face electrodes (Ashimbaeva et al., 2004). Based on the mathematical model outlined here, we now consider further the focusing properties of that analyzer.

In the coordinate system r, z, the hexapole-cylindrical field at $\gamma=-1$ appears as

$$U(r, z) = \ln r - U_h(r, z). \tag{30}$$

Such a superposition of the cylindrical and hexapole fields can be implemented in an axisymmetric electrostatic mirror-type analyzer, the schematic of which is shown in Figure 3.The analyzer comprises a cylindrical mirror ($r_0=1$), two end-face electrodes, and an axisymmetric deflecting electrode, the generatrix of which coincides with one of the equipotential lines. The cylindrical and end-face electrodes are at ground potential, while the sign of the potential U_o of the electrode with a curved profile coincides with the sign of the particle's charge.

In coordinates x and ξ, the distribution [Eq. (30)] reads as

$$U(x, \xi) = U_o\, g(x, \xi), \tag{31}$$

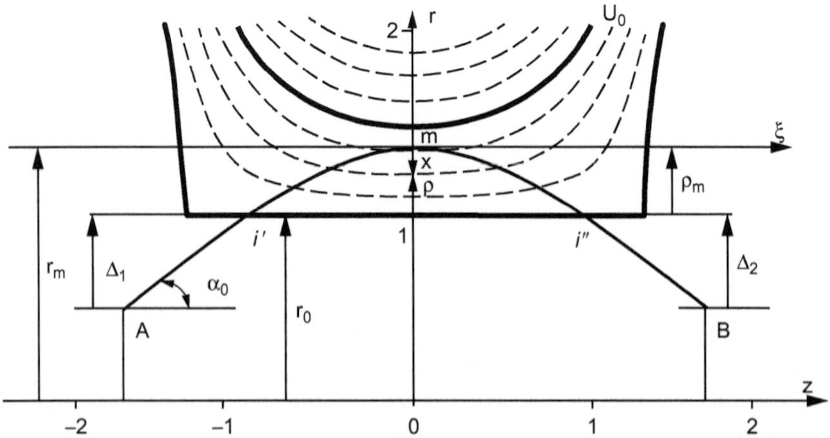

Figure 3 A scheme of the mirror-type hexapole-cylindrical analyzer with end-face electrodes (A—source, i' —input annular slit, i'' —output annular slit, B—receiver).

where

$$g(x,\xi) = \ln(R-x)\left[-\frac{1}{2}\xi^2 + \frac{1}{4}\left(R-x)^2\right) + \frac{5}{4}\right] - \frac{1}{4}(R-x)^2) + \frac{1}{4},$$
$$R = 1 + \rho_m. \tag{32}$$

Let us represent the motion equation of charged particles in the field [Eq. (32)] as

$$m\ddot{x} = qU_0\varepsilon_1, \quad \varepsilon_1 = -\frac{\partial g(x,\xi)}{\partial x}, \tag{33}$$

$$m\ddot{\xi} = qU_0\varepsilon_2, \quad \varepsilon_2 = -\frac{\partial g(x,\xi)}{\partial \xi}. \tag{34}$$

By using the transformations we have applied already in this chapter, the system of Eqs. (33) and (34) can be transformed into the integro-differential equation [Eq. (17)] describing the charged particle motion along a "return" trajectory in the hexapole-cylindrical field [Eq. (32)], with

$$g_o = g(x_m, \xi_m) = \ln(R)\left[\frac{R^2}{4} + \frac{5}{4}\right] - \frac{R^2}{4} + \frac{1}{4},$$
$$g_x = g(x, \xi(x)). \tag{35}$$

To solve Eq. (17) and avoid the uncertainties at the singular point $x=0$ when calculating ξ', we apply the method of expansion to the fractional-power series [Eq. (21)].

The radial component of the trajectory's turning point $R=1+\rho_m$ can be calculated from Eq. (17) using the relations $x=\rho_m$, $(\xi')^2=ctg^2\alpha_0$, $g_{x=\rho_m}=0$ and the value g_o determined from Eq. (35):

$$\ln R = \frac{4(P^2+f_m)+R^2-1}{R^2+5}. \tag{36}$$

This transcendent equation can be solved using the method of consequent approximations, with the initial approximation representing the parameter $R_0=\exp\left(P^2\right)$ of the cylindrical mirror-type analyzer (Zashkvara et al., 1966).

Next, we present the final results of calculations, in which the accuracy of determining ξ is limited to the value $c_6\,x^{13/2}$, while the parameter $\rho_m = R-1$ is calculated in the form of expansion into a series by the reflection parameter P with an accuracy up to the 14th-order inclusive:

$$\begin{aligned}\rho_m = {} & P^2+0.5P^4+\mathrm{ctg}^2(\alpha_o)P^6+\left(2.55556\mathrm{ctg}^2(\alpha_o)-0{,}29167\right)P^8\\ & +\left(2.22222\mathrm{ctg}^4(\alpha_o)+1{,}42778\mathrm{ctg}^2(\alpha_o)-0{,}25833\right)P^{10}\\ & +\left(11.18889\mathrm{ctg}^4(\alpha_o)-4{,}89317\mathrm{ctg}^2(\alpha_o)+0{,}03750\right)P^{12}\\ & +\left(6.22222\mathrm{ctg}^6(\alpha_o)-16{,}77238\mathrm{ctg}^4(\alpha_o)-11.44404\mathrm{ctg}^2(\alpha_o)+0{,}32182\right)P^{14}\ldots\end{aligned} \tag{37}$$

$$\xi_m = \xi(x)\big|_{x=\rho_m} = \sqrt{\rho_m}\,C_o S,$$

where

$$C_o = ctg\alpha_o\sqrt{\begin{aligned}&4P^2+4P^4-4P^6+\left(4.0\mathrm{ctg}^2\alpha_o-14.88889\right)P^8\\&+\left(-2.66667\mathrm{ctg}^2\alpha_o-10.87778\right)P^{10}\\&+\left(8.88889\mathrm{ctg}^4\alpha_o-61.26667\mathrm{ctg}^2\alpha+20.47270\right)P^{12}\ldots\end{aligned}}, \tag{38}$$

$$\begin{aligned}S = {} & 1-0.08333P^2+\left(0.16667\mathrm{ctg}^2\alpha_o+0.32708\right)P^4\\ & +\left(0.14722\mathrm{ctg}^2\alpha_o+0.42607\right)P^6\\ & +\left(0.26389\mathrm{ctg}^4\alpha_o+0.53897\mathrm{ctg}^2\alpha_o+0.13707\right)P^8\\ & +\left(0.81703\mathrm{ctg}^4\alpha_o+1.93841\mathrm{ctg}^2\alpha_o-0.36304\right)P^{10}\\ & +\left(0.64120\mathrm{ctg}^6\alpha_o+1.97321\mathrm{ctg}^4\alpha_o+2.26796\mathrm{ctg}^2\alpha_o-0.58157\right)P^{12}\ldots\end{aligned} \tag{39}$$

To implement the second-order angular focusing at a certain distance l with respect to the beam's divergence angle $\Delta\alpha$ in the axial plane of the mirror, the condition $\frac{dl}{d\alpha} = \frac{d^2 l}{d\alpha^2} = 0$ must be fulfilled. Table 1 in Appendix 2 represents the basic electron-optical parameters of the analyzer, calculated as functions of the reflection parameter P [see Eq. (18)] and satisfying the conditions of second–order angular focusing. Here. $\alpha_{\circ}$ is the tilting angle of the axial trajectory at the entrance into the mirror field; $\Delta = \Delta_1 + \Delta_2$ is the total distance of the source and its image from the interior cylindrical electrode; ρ_m is the coordinate of the trajectory's turning point, which determines the value of maximal passing of a particle into the mirror field; $\boldsymbol{l}$ is the focal distance equal to the full projection of the charged particle trajectory onto the symmetry axis of the mirror from the source to its image; $D = \frac{\partial l}{\partial \varepsilon}$ is the relative linear dispersion in energy determined by differentiation of l in accordance with the energy spread of particles $\varepsilon = \frac{\Delta\omega}{\omega}$; $A_{III} = \frac{1}{3!}\frac{d^3 l}{d\alpha^3}$ is the coefficient of third-order angular aberration; and $\delta = \frac{D}{A_{III} \cdot (\Delta\alpha)^3}$ is the specific energy dispersion that characterizes the device's resolution.

Numerical analysis of the characteristics of the mirror-type analyzer based on the hexapole-cylindrical field shows that the mode of second-order angular focusing can be implemented in a small interval of input angles $\alpha_o = (30.0 \div 3.4)^\circ$, and only for the schematics with $\Delta > 0$, when the source and its image are located in the interior cylinder region. The increase in P leads to the gradual increase in all basic mirror parameters, with the exception of α_o and specific dispersion δ that is increased up to a certain maximal value in the region $P = 0.25$–0.3 and then falls to a minimum in the region $P > 0.5$; so further calculations in the region of greater P make no practical sense (Ashimbaeva et al., 2004).

Table 1 indicates that the hexapole-cylindrical analyzer under consideration allows implementation of the well-known schemes of angular focusing within the"ring-axis" scheme for $\Delta = 1$, when an annular source is imaged into a point on the symmetry axis, and within the "axis-axis" scheme for $\Delta = 2$, wherein the mode of focusing of a point source into a point image is realized. Despite the existence of electron-optical schemes with a higher resolution in the region of small values of the reflection parameter P, the focusing scheme of "ring-axis" type, with $P = 0.4892$, and $\Delta = 1$, may be considered as the most optimal one. This scheme possesses the greatest value of linear dispersion by energy, which characterizes the source image's displacement as a function of the charged particle's energy.

The main advantage of the mirror analyzer proposed is the presence of two end-face electrodes that, along with the interior cylindrical electrode, are placed at ground potential. The end-face electrodes restrict the longitudinal dimensions of the analyzer and minimize the fringe field effect on the distribution of a retarding field in the region of charged particle motion. The calculations show that the distance along the interior cylinder surface in the units of r_o is $2\sqrt{2}$. At small ρ_m values, the generatrix of the end-face electrodes, which lies in the axial plane, is close to a straight line with a slight deviation from the mirror radial plane. In particular, in the case of "ring-axis" focusing with $P=0.4892$, the generatrix's tilting angle to the symmetry axis in the region $\rho_m \leq 0.4$ constitutes 87.3°, which means that the end-face electrode surface can be approximated by a conical surface with a slight deviation from the radial plane.

It is worthwhile to pay attention to one more advantage of the hexapole-cylindrical analyzers. The issue is that a classic cylindrical mirror-type analyzer (Zashkvara et al., 1966) cannot in principle provide a long focus. In recent decades, several new designs of axisymmetric energy analyzers have been proposed as an alternative to the analyzer based on the cylindrical mirror. The quasi-conical energy analyzer (Siegbahn, Kholine, & Golikov, 1997), despite having a number of obvious advantages (in particular, much higher energy resolution compared to the cylindrical mirror), has a too small "sample-analyzer" distance, which restricts its use in real devices. That problem can be solved by applying the electrostatic hexapole-cylindrical fields proposed here.

In seeking a schematic with angular focusing of the "axis-ring" type that ensures a large focus distance from the sample, we performed a numerical modeling of the hexapole-cylindrical analyzer with protection from the fringe fields. The calculations have been performed using the FOCUS software designated for the modeling of electron-optical systems (Trubitsyn, 2008).

Figure 4 shows a set of charged particle trajectories emitted from a point source in the 25°–35° range of input angles in a long-focus hexapole-cylindrical analyzer at $\gamma = -1$. The source of electrons represents an electron gun, the axis of which coincides with the symmetry axis of the energy analyzer.

Figure 4 demonstrates how the secondary electrons (labeled 4 in the figure) excited in a test sample (5) by means of the primary radiation (6; i.e., electrons) pass through a special entrance window (8) in the interior cylinder covered by a one-dimensional (1D) grid into a hexapole-cylindrical field

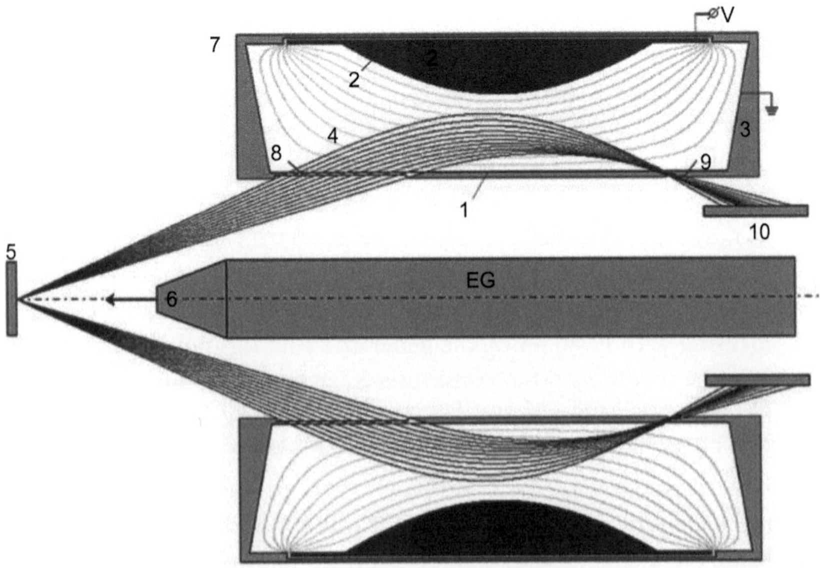

Figure 4 Electron-optical scheme of the long-focus hexapole-cylindrical analyzer providing the mode of axis-ring angular focusing (1–cylindrical electrode, 2–outer electrode with a curved profile, 3–end-face electrodes, 4–secondary electrons, 5–sample, 6–primary electrons, 7–magnetic screen, 8 and 9–entrance and exit slits, 10–detector, EG–electron gun). (See the color plate.)

that deviates the particles toward the cylinder's axis and focuses them onto the cylindrical electrode surface. The electrons that have passed into the inner cylinder (1) through the output window's grid (9) are registered by the detector (10). The scheme operates in the mode of second-order angular focusing of the axis-ring type. The input and output slits are covered by fine-structure grids that scatter the bunch only in an azimuth direction.

The electron-optical scheme modeling has shown the ability to implement the mode of second-order focusing near the 30° angle of particle emission from the source. Such a scheme provides an efficient mode of transportation onto the detector of the particle flow in the range of emission angles of $30° \pm 5°$, thus solving the problem of increasing the device's sensitivity. Calculations also indicate that the energy analyzer based on the superposition of the circular cylindrical field and the hexapole with $\gamma = -1$ has significant advantages compared to traditional types of energy analyzers. High luminosity, high energy resolution, large sample-analyzer distance, and relatively easy implementation of the curved outer electrode profile are important characteristics of this energy analyzer.

These features allow the use of the proposed analyzer not only in the Auger electron spectroscopy, but also in photoelectron spectroscopy with different excitation modes. The presence of the analyzer's end-face electrodes restricting the longitudinal dimension of the analyzer ensures minimization of the fringe field impact on the basic field distribution.

4. ELECTRON-OPTICAL PROPERTIES OF THE HEXAPOLE-CYLINDRICAL ENERGY ANALYZER WITH $\gamma = 1$

This section discusses the analysis of the trajectories and electron-optical characteristics for the case of charged particle motion in an electrostatic field, the potential of which, in the coordinate system r, z, is described by the expression

$$U(r, z) = \ln r + U_h(r, z). \quad (40)$$

A schematic view of the energy analyzer with such a field distribution is shown in Figure 5 (Ashimbaeva, Chokin, Saulebekov, & Kambarova, 2012b). The hexapole-cylindrical field is formed in the space between the interior of the cylindrical electrode placed at ground potential and the

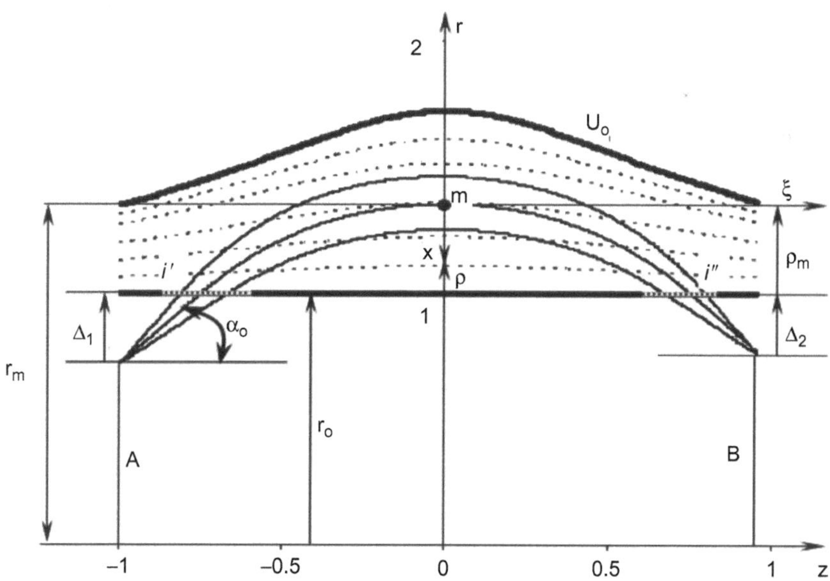

Figure 5 Schematic view of the energy analyzer based on the hexapole-cylindrical field with $\gamma = 1$ (A—source, i' —input annular slit, i'' —output annular slit, B—detector).

outer electrode with a curved profile, carrying a deflecting potential U_0. According to the scheme, a bunch of charged particles leaving the annular source A is reflected in the analyzer's field and focused into annular image B.

An annular source and annular image are often required for structural reasons. An annular electron-optical source (e.g., an annular diaphragm being irradiated by a flow of particles coming from a sample of a large area) is used for the excitation of electrons by X-rays, in scanning Auger microscopy, and other methods.

For subsequent calculations of the electron-optical scheme, we will use the procedure given earlier in this chapter [Eqs. (31)–(35)]. Let us first transform the origin point of the particle trajectory in the field to the trajectory vertex m. Thus, the coordinate system origin is located at the point x, ξ (Figure 5). All linear dimensions in the calculations are normalized to the inner cylindrical electrode radius r_0 [Eq. (3)].

In the new coordinate system x, ξ, the potential distribution given by Eq. (40) takes the form of Eq. (31), while Eq. (32) appears as

$$g(x,\xi)=\ln(R-x)\left[\frac{1}{2}\xi^2-\frac{1}{4}\left(R-x)^2\right)+\frac{3}{4}\right]+\frac{1}{4}\left(R-x)^2\right)-\frac{1}{4},$$
$$R=1+\rho_m. \tag{41}$$

The equations of the charged particle motion in the field [Eqs. (40) and (41)] retain the view of Eqs. (33) and (34). Integrating of the sum of these equations along the particle trajectory in the region from the trajectory vertex m to an arbitrary point gives the law of energy conservation for the particle motion in the electrostatic field under study. This expression relates the change in kinetic energy of a particle with the potential difference the particle has passed:

$$\frac{mv_m^2}{2}-\frac{m}{2}\left(\dot{x}^2+\dot{\xi}^2\right)=-q(U_m-U)=-qU_0(g_0-g_x). \tag{42}$$

where $U_m=U_0 g(x_m,\xi_m) = U_0 g_0$ is the potential at point m, at which $x_m=\xi_m=0$, $g_x=g(x,\xi(x))$.

The value $\frac{m\dot{\xi}^2}{2}$ can be determined by integration of the second part of Eq. (34) in the region from m to an arbitrary point on the particle trajectory. Herewith, we take into consideration that $v_m^2=\dot{\xi}_m^2+\dot{x}_m^2 = \dot{\xi}_m^2$, since at the trajectory vertex, $\dot{x}_m = 0$.

Using the relation $\dot{\xi} = \frac{d\xi}{dt} = \frac{d\xi}{dx}\frac{dx}{dt} = \xi' \dot{x}$, we obtain

$$\frac{mv_m^2}{2} - \frac{m\dot{\xi}^2}{2} = q\,U_0 \int\limits_{x}^{x_m} \frac{\partial g(x,\xi)}{\partial \xi}\,\frac{d\xi}{dx}\,dx = q\,U_0 \int\limits_{0}^{x} \ln(R-x)\,\xi\xi'\,d\xi. \tag{43}$$

According to Figure 2, $\frac{m\dot{\xi}^2}{2} = W\cos^2\alpha_0$ at $x = \rho_m$, where W is the kinetic energy of a particle. Therefore, Eq. (43) can be rewritten as

$$\frac{mv_m^2}{2} = W\cos^2\alpha + qU_0 f_m,$$
$$f_m = \int\limits_{0}^{\rho_m} \frac{\partial g(x,\xi)}{\partial \xi}\,\frac{d\xi}{dx}\,dx = \int\limits_{0}^{\rho_m} \ln(R-x)\,\xi\xi'\,d\xi. \tag{44}$$

Substituting the right-hand sides of Eqs. (43) and (44) into Eq. (42), we obtain an integro-differential equation of the charged particle motion in the hexapole-cylindrical field with $\gamma = 1$:

$$(\xi')^2\left[g_0 - g_x + \int\limits_{0}^{x} \ln(R-x)\,\xi\xi'\,dx\right] = P^2\,\mathrm{ctg}^2\alpha_0 + f_m - \int\limits_{0}^{x} \ln(R-x)\,\xi\xi'\,dx, \tag{45}$$

where

$$g_0 = g(x_m, \xi_m) = \ln R\left[\frac{R^2}{4} + \frac{3}{4}\right] - \frac{R^2}{4} + \frac{1}{4}. \tag{46}$$

And $P_0^2 = \frac{W}{qU}\sin^2\alpha_0$ is the reflection parameter of the hexapole–cylindrical analyzer, interrelating its geometric and energetic parameters.

In order to solve the integro-differential equation [Eq. (45)], the method of power series expansions is used. The sought function in Eq. (45) has a discontinuity at the singular point $x = 0$, with the factor $(\xi')^2$ vanishing. This function cannot be represented in the form of a conventional power series; therefore, the solution is sought in the form of a generalized series $\xi = x^{\rho}\sum_{n=0}^{\infty} C_n x^n$. We easily obtain $\rho = {}^1/_2$, and accordingly,

$$\xi = \sqrt{x}\left(C_0 + C_1x + C_2x^2 + C_3x^3 + C_4x^4 + C_5x^5 + C_6x^6 + \ldots\right). \tag{47}$$

Next, we represent the coefficient C_n calculated by substituting the fractional-power series [Eq. (47)] into Eq.(45):

$$C_0^2 = 4\frac{P^2 \operatorname{ctg}^2\alpha_0 - f_m}{\gamma}, \gamma = \frac{1}{4}\left(R + \frac{3}{R} - 2R\ln R\right), \tag{48}$$

$$\frac{C_1}{C_0} = \frac{1}{\gamma}\left[-\frac{C_0^2}{24R} - \frac{3}{8}\ln R - \frac{11}{16R^2} - \frac{1}{48}\right], \tag{49}$$

$$\begin{aligned}\frac{C_2}{C_0} = &-\frac{9}{10}\frac{C_1^2}{C_0^2} + \frac{1}{\gamma} \times \left[\frac{C_1}{C_0}\left(-\frac{11}{60}\frac{C_0^2}{R} - \frac{11}{20}\ln R - \frac{9}{40R^2} - \frac{3}{40}\right)\right. \\ &\left. - \frac{1}{60}\frac{C_0^2}{R^2} - \frac{1}{40R^3} + \frac{13}{120R}\right],\end{aligned} \tag{50}$$

$$\begin{aligned}\frac{C_3}{C_0} = &-\frac{15}{7}\frac{C_1 C_2}{C_0^2} + \frac{1}{\gamma}\left[\frac{C_2}{C_0}\left(-\frac{11}{56}\frac{C_0^2}{R} - \frac{13}{28}\ln R - \frac{15}{56R^2} - \frac{5}{56}\right)\right. \\ &+ \frac{C_1^2}{C_0^2}\left(-\frac{5}{16}\frac{C_0^2}{R} - \frac{17}{56}\ln R - \frac{27}{112R^2} - \frac{9}{112}\right) \\ &\left. + \frac{C_1}{C_0}\left(-\frac{5}{56}\frac{C_0^2}{R} - \frac{3}{28R^3} + \frac{19}{84R}\right) - \frac{C_0^2}{112R^3} - \frac{3}{224R^4} + \frac{17}{672R^2}\right],\end{aligned} \tag{51}$$

Eqs. (48)–(51) imply that the coefficient C_n needed for calculating the function $\xi(x) = \sqrt{x}\sum_{n=0}^{\infty} C_n x^n$ depends on the radial coordinate R_0 of the trajectory's turning point. This value can be found from Eq. (45) if $x = \rho_m$. So we obtain $(\xi')^2 = \operatorname{ctg}^2\alpha_0$, $g_{x=\rho_m} = 0$, and

$$g_0 + f_m = P^2. \tag{52}$$

Substituting g_0 determined by Eq. (46) into Eq. (52), we come to the transcendent equation

$$\ln R = \frac{4(P^2 - f_m) - R^2 + 1}{3 - R^2}, \tag{53}$$

from which the value of R can be found by the method of consecutive approximations, with the parameters of the cylindrical mirror-type analyzer taken as zero approximation (Zashkvara et al., 1966):

$$R_0 = \exp P^2 = 1 + P^2 + \frac{1}{2}P^4 + \frac{1}{6}P^6 + \frac{1}{24}P^8 + \ldots$$

and

$$f_{m_0} = 0.$$

Next, the results of the particle trajectory calculation in the hexapole-cylindrical field are represented in the form of a series expansion by the reflection parameter P up to the 14th order inclusive:

$$\rho_m = P^2 + \frac{1}{2}P^4 + \left[-\mathrm{ctg}^2\alpha_0 + \frac{1}{3}\right]P^6 + \left[-\frac{23}{9}\mathrm{ctg}^2\alpha_0 + \frac{3}{8}\right]P^8$$
$$+ \left[\frac{20}{9}\mathrm{ctg}^4\alpha_0 - \frac{313}{60}\mathrm{ctg}^2\alpha_0 + \frac{53}{120}\right]P^{10} \tag{54}$$
$$+ \left[\frac{1007}{90}\mathrm{ctg}^4\alpha_0 - \frac{67631}{6300}\mathrm{ctg}^4\alpha_0 + \frac{79}{144}\right]P^{12}\ldots$$

The projection ξ_m onto the mirror's symmetry axis of the trajectory between the particle's entry point into the field and the turning point m is given by Eq. (47), provided that $x = \rho_m$:

$$\xi_m = \xi(x)_{|x=\rho_m} = \sqrt{\rho_m}\, C_0\, S, \tag{55}$$

where

$$C_0 = \mathrm{ctg}\alpha_0 \sqrt{\begin{aligned} &4P^2 + 4P^4 + 8P^6 + \left(-4\mathrm{ctg}^2\alpha_0 + \frac{146}{9}\right)P^8 \\ &+ \left(-\frac{208}{9}\mathrm{ctg}^2\alpha_0 + \frac{297}{10}\right)P^{10} \\ &+ \left(\frac{80}{9}\mathrm{ctg}^4\alpha_0 - \frac{3953}{45}\mathrm{ctg}^2\alpha_0 + \frac{176737}{3150}\right)P^{12}\ldots \end{aligned}}, \tag{56}$$

$$S = 1 + \frac{C_1}{C_o}\rho_m + \frac{C_2}{C_o}\rho_m^2 + \frac{C_3}{C_o}\rho_m^3 + \frac{C_4}{C_o}\rho_m^4 + \frac{C_5}{C_o}\rho_m^5 + \frac{C_6}{C_o}\rho_m^6\ldots$$
$$= 1 - \frac{1}{12}P^2 + \left(-\frac{1}{6}\mathrm{ctg}^2\alpha_0 - \frac{131}{480}\right)P^4 + \left(-\frac{53}{360}\mathrm{ctg}^2\alpha_0 - \frac{17573}{40320}\right)P^6 \tag{57}$$
$$+ \left(0,26389\,\mathrm{ctg}^4\alpha_0 + 0,34595\,\mathrm{ctg}^2\alpha_0 - 0,50249\right)P^8$$
$$+ \left(0,81703\,\mathrm{ctg}^4\alpha_0 + 1,83040\,\mathrm{ctg}^2\alpha_0 - 0,70059\right)P^{10}\ldots$$

A full projection of the trajectory onto the symmetry axis from the source A up to its image B represents the sum

$$l = \frac{l}{r_0} = \Delta \mathrm{ctg}\alpha_0 + 2\xi_m, \ \Delta = \Delta_1 + \Delta_2, \tag{58}$$

where Δ_1 and Δ_2 are the distances of the source and its image up to the interior cylindrical electrode surface (Figure 5).

The expression describing the charged particle trajectories in that hexapole-cylindrical field determines all necessary electron-optical characteristics of the analyzer. To analyze those parameters, the angular focusing coefficients up to the third order inclusive (namely, $A_I = \frac{dl}{d\alpha}$, $A_{II} = \frac{1}{2!}\frac{d^2l}{d\alpha^2}$, and $A_{III} = \frac{1}{3!}\frac{d^3l}{d\alpha^3}$) must be calculated. The condition of second-order focusing appears as $\frac{dl}{d\alpha} = \frac{d^2l}{d\alpha^2} = 0$.

Table 2 of Appendix 2 shows the main electron-optical characteristic of the analyzer as functions of the reflection parameter P and the angle α_0 of entry of the trajectories into the field, with the condition of second-order focusing fulfilled. All these parameters have been described earlier in this chapter when analyzing the data of Table 1 except the parameter ξ_m, which represents half of the trajectory projection onto the symmetry axis in the mirror field.

Analysis of the data in Table 2 hinges on the fact that the hexapole-cylindrical mirror energy analyzer allows the implementation of the mode of second-order focusing in a wide range of the values of parameter P and the input angle. Herewith, the source and its image are located in the region of the inner cylindrical electrode, which means that the condition of angular focusing can only be realized for the schemes of focusing the annular source into the annular detector ($\Delta_1 + \Delta_2 < 1$).

The sole scheme of an analyzer, which allows focusing in the "ring-axis" mode with the parameters $P = 0.7045$, $\alpha_o = 44.86°$, and $\Delta_1 + \Delta_2 = 1$, is of no practical interest. In this case, the coefficient of cubic angular aberration of the device is quite large ($A_{III} \sim 33.16$). The most optimal design of the mirror analyzer with the hexapole-cylindrical field is realized in the region $0.52 < P < 0.64$, for which the third-order aberration coefficients are small.

Defined here are two schemes with the parameters $P = 0.5754$, $\alpha_0 = 40.9766°$, and $P = 0.6242$, $\alpha_0 = 46.5065°$ that ensure the mode of third-order angular focusing $\frac{dl}{d\alpha} = \frac{d^2l}{d\alpha^2} = \frac{d^3l}{d\alpha^3} = 0$. However, the greatest interest lies in the electron-optical scheme of the hexapole-cylindrical analyzer with the parameters $P = 0.6$, $\alpha_0 = 43.7420°$, $D = 1.7967$, $A_{III} = 0.7340$. Cubic aberration in this scheme is small and attains a maximum in the region

$0.52 < P < 0.64$. This implies that the next-order aberrational coefficient A_{IV} vanishes at that point, which allows us to conclude that the angular focusing in this analyzer is close to ideal (Ashimbaeva et al., 2012b).

The entry angle of the axial trajectory into the mirror field is $\alpha_0 = 43.7420°$, while the initial angular spread for the side branches of the trajectory amounts to $\Delta\alpha = \pm 8°$. The image blur at the energy analyzer focus, caused by the beam's angular divergence of 16° in the axial plane, is determined by the cubic angular aberration $\Delta l = A_{III}(\Delta\alpha)^3$ and constitutes $\Delta l = 0{,}004$. This means that the energy analyzer with such a field distribution is capable of the sharp focusing of charged particles with a large energy spread in the axial plane. For comparison, the line width of the image of a particle beam with the same angular spread in the cylindrical mirror analyzer (Zashkvara et al., 1966) is approximately 20 times larger and constitutes $\Delta l = 0{,}084$.

If the annular slits i' and i" are narrow (Figure 5), and the beam's angular divergence is large, the analyzer's resolution can be estimated by the magnitude of specific energy dispersion (Kozlov, 1978). This value is equal to the ratio of the linear dispersion by energy to the line image width in the mirror focus:

$$\delta = \frac{D}{\Delta l} = \frac{D}{A_{III}\,(\Delta\alpha)^3}. \tag{59}$$

The specific dispersion in energy of the analyzer for the particles with an initial angular spread of 16° amounts to $\delta = 449{,}68$. This value is approximately seven times higher than the relevant specific dispersion in a cylindrical mirror.

In order to confirm the results obtained by the approximate-analytical method, numerical simulation of electron–optical scheme of the analyzer based on the hexapole-cylindrical field at $\gamma = 1$ was performed using the FOCUS software (Trubitsyn, 2008).

According to Figure 6, the charged particles emitted from a thin annular source S in the range of polar angles from 31° to 44° are deflected under the action of the potential on the electrode with a curved exterior profile (labeled 2 in the figure) and then focused into an annular image S'. The annular source S and its image S' are located in the area of the interior cylindrical electrode. Thus, the system operates in the angular focusing mode of the "ring-ring" type.

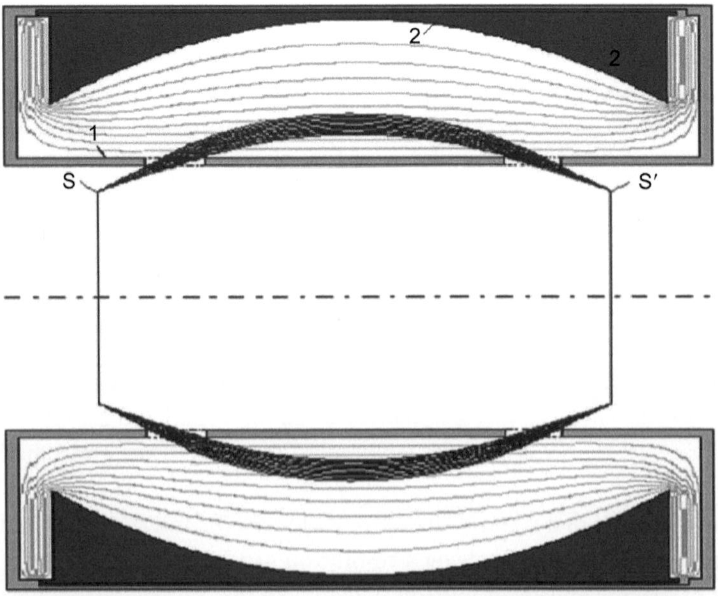

Figure 6 Charged particle trajectories in the hexapole-cylindrical analyzer with $\gamma = 1$ (1—cylindrical electrode, 2—external deflecting electrode with a curved profile, *S*—annular source, *S*′—annular image). (See the color plate.)

Our analysis shows that the electron-optical system under consideration allows implementation of the mode of second-order angular focusing in a wide range of input angles (31°–44°) with a central angle of 35.70°. For this system, two schemes are found, providing the mode of third-order angular focusing.

Third-order focusing is found in the range of input angles of 39°–42° with a central angle of 40.95° and in the range of input angles of 45°–48° with a central angle of 46.55°. The results derived by using the approximate-analytical method are in good agreement with the results of numerical simulations.

Thus, we have calculated the electron-optical parameters of the mirror-type energy analyzer based on the electrostatic hexapole-cylindrical field with $\gamma = 1$ and established the conditions of angular focusing of charged particle beams with a large angular divergence in the axial plane. It was found that if the source and its image are located in the region of the inner cylindrical electrode, the field distributions under study may constitute a basis for designing the energy analyzers with second- and third-order angular focusing. Determined are the electron-optical parameters of the hexapole-cylindrical mirror-type energy analyzer with a close-to-ideal angular focusing that ensures high-resolution and large-luminosity operation modes.

5. ANALYSIS OF ELECTRON-OPTICAL CHARACTERISTICS OF THE ENERGY ANALYZER WITH THE FIELD DISTRIBUTION $U\,(r,z) = \frac{5}{2}\ \ln r - U_h\,(r,z)$

Consider the charged particle motion in a hexapole-cylindrical field in the case that the coefficients characterizing the partial contribution of the cylindrical and hexapole fields are equal to 5/2 and −1, respectively (Ashimbaeva, Chokin, & Saulebekov, 2003). In this case, the potential distribution in the coordinate system *r, z* appears as

$$U(r,z) = \frac{5}{2}\ln r - U_h(r,z). \tag{60}$$

A schematic view of the relevant energy analyzer is shown in Figure 7. At a certain ratio of geometric and energy parameters, the charged particle beam coming from the annular source is reflected by the mirror field and focused into an annular image.

In terms of the coordinates x and ξ, the field distribution [Eq. (60)] acquires the form [Eq. (10)] with

$$g(x,\xi) = \ln(R-x)\left[-\frac{1}{2}\xi^2 + \frac{1}{4}(R-x)^2 + \frac{11}{4}\right] - \frac{1}{4}(R-x)^2 + \frac{1}{4},$$
$$R = 1 + \rho_m. \tag{61}$$

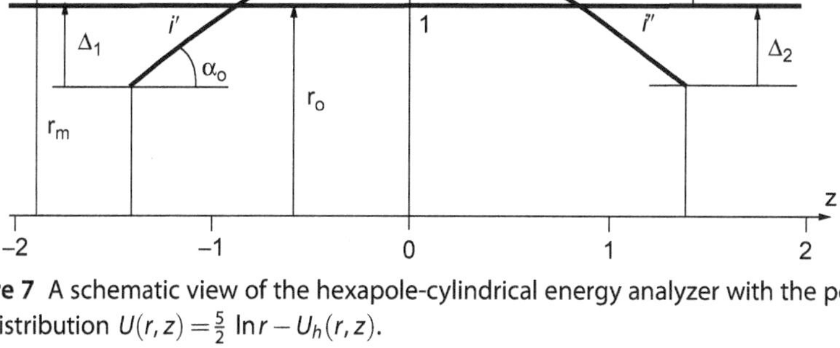

Figure 7 A schematic view of the hexapole-cylindrical energy analyzer with the potential distribution $U(r,z) = \frac{5}{2}\ln r - U_h(r,z)$.

In accordance with the arguments set out in section 1, we obtain the integral-differential equation of charged particle motion along the return trajectory in the field [Eq. (60)]:

$$(\xi')^2\left[g_o - g_x - \int_0^x \ln(R-x)\,\xi\xi'\,dx\right] = P^2\,\mathrm{ctg}^2\alpha_o - f_m + \int_0^x \ln(R-x)\,\xi\xi'\,dx, \tag{62}$$

where

$$g_o = g(x_m, \xi_m)\Big|_{\substack{x_m=0\\ \xi_m=0}} = \ln(R)\left[\frac{R^2}{4}+\frac{11}{4}\right] - \frac{R^2}{4}+\frac{1}{4},\ g_x = g(x,\xi(x)). \tag{63}$$

To solve the integral-differential equation [Eq. (62)], we again use the method of the power series expansions [Eq. (21)]. In this case, the transcendent equation with respect to R takes the form

$$\ln R = \frac{4(P^2+f_m)+R^2-1}{R^2+11}, \tag{64}$$

from which R can be determined by means of the method of successive approximations. As before, the parameters of the cylindrical mirror-type energy analyzer are used as a zero approximation.

Next, we present the final results of calculating the parameters $\rho_m = R-1$ and C_m, which are necessary for calculating the trajectory projection onto the symmetry axis in the hexapole-cylindrical mirror. All calculations are made in the form of power series expansions with respect to the mirror reflection parameter P:

$$\begin{aligned}\rho_m = {} & 0.4P^2 + 0.08P^4 + \left(0.0256\,\mathrm{ctg}^2(\alpha_o) - 0.0192\right)P^6\\ & + \left(0.0262\,\mathrm{ctg}^2(\alpha_o) - 0.0285\right)P^8\\ & + \left(0.0036\,\mathrm{ctg}^4(\alpha_o) + 0.0037\,\mathrm{ctg}^2(\alpha_o) - 0.0085\right)P^{10}\ldots\end{aligned} \tag{65}$$

$$\begin{aligned}C_o^2 = {} & \left(1.6\,\mathrm{ctg}^2(\alpha_o) - 1.6\right)P^2 + \left(0.64\,\mathrm{ctg}^2(\alpha_o) - 0.64\right)P^4\\ & + \left(-0.0256\,\mathrm{ctg}^2(\alpha_o) + 0.0256\right)P^6\\ & + \left(0.0410\,\mathrm{ctg}^4(\alpha_o) - 0.2241\,\mathrm{ctg}^2(\alpha_o) + 0.1832\right)P^8\\ & + \left(0.0208\,\mathrm{ctg}^4(\alpha_o) - 0.1096\,\mathrm{ctg}^2(\alpha_o) + 0.0888\right)P^{10}\ldots\end{aligned} \tag{66}$$

$$\frac{C_1}{C_o}=-0.0833+\left(0.0267\text{ctg}^2(\alpha_o)+0.0733\right)P^2$$
$$+\left(0.0107\text{ctg}^2(\alpha_o)+0.0093\right)P^4+\left(0.0051\text{ctg}^2(\alpha_o)-0.0021\right)P^6 \quad (67)$$
$$+\left(0.0007\text{ctg}^4(\alpha_o)+0.0020\text{ctg}^2(\alpha_o)-0.0050\right)P^8$$
$$+\left(0.0011\text{ctg}^4(\alpha_o)-0.0007\text{ctg}^2(\alpha_o)-0.0017\right)P^{10}\ldots$$

$$\frac{C_2}{C_o}=0.0488+\left(-0.0381\text{ctg}^2(\alpha_o)-0.1427\right)P^2$$
$$+\left(0.0025\text{ctg}^4(\alpha_o)+0.0026\text{ctg}^2(\alpha_o)+0.0269\right)P^4$$
$$+\left(0.0020\text{ctg}^4(\alpha_o)-0.0034\text{ctg}^2(\alpha_o)+0.0059\right)P^6$$
$$+\left(0.0006\text{ctg}^4(\alpha_o)+0.0002\text{ctg}^2(\alpha_o)+0.0055\right)P^8\ldots \quad (68)$$

$$\frac{C_3}{C_o}=-0.0201+\left(0.0074\text{ctg}^2(\alpha_o)+0.0257\right)P^2$$
$$+\left(-0.0027\text{ctg}^4(\alpha_o)-0.0090\text{ctg}^2(\alpha_o)-0.0096\right)P^4 \quad (69)$$
$$+\left(0.0003\text{ctg}^6(\alpha_o)-0.0008\text{ctg}^4(\alpha_o)+0.0007\text{ctg}^2(\alpha_o)+0.0073\right)P^6\ldots$$

$$\frac{C_4}{C_o}=-0.0181+\left(0.0105\text{ctg}^2(\alpha_o)+0.0547\right)P^2$$
$$+\left(-0.0013\text{ctg}^4(\alpha_o)-0.0153\text{ctg}^2(\alpha_o)-0.0484\right)P^4 \quad (70)$$
$$+\left(-0.0004\text{ctg}^6(\alpha_o)-0.001\text{ctg}^4(\alpha_o)+0.0063\text{ctg}^2(\alpha_o)+0.0155\right)P^6\ldots$$

$$\frac{C_5}{C_o}=\left(-0.0023\text{ctg}^2(\alpha_o)+0.0019\right)P^2$$
$$+\left(0.0009\text{ctg}^4(\alpha_o)+0.0027\text{ctg}^2(\alpha_o)-0.0072\right)P^4\ldots \quad (71)$$

$$\frac{C_6}{C_o}=0.0023+\left(-0.0043\text{ctg}^2(\alpha_o)-0.0154\right)P^2\ldots. \quad (72)$$

Based on Eqs. (65)–(72), it is possible to determine the trajectory projection in the mirror field from the entry point to the return point m in the form of an expansion into a series by the value $x=\rho_m$:

$$\xi_m=\xi_{|x=\rho_m}=\sqrt{\rho_m C_o^2}S, \quad (73)$$

where

$$S=1+\frac{C_1}{C_o}\rho_m+\frac{C_2}{C_o}\rho_m^2+\frac{C_3}{C_o}\rho_m^3+\frac{C_4}{C_o}\rho_m^4+\frac{C_5}{C_o}\rho_m^5+\frac{C_6}{C_o}\rho_m^6\ldots$$

$$\begin{aligned}&= 1. - 0.0333P^2 + \left(0.0107\mathrm{ctg}^2(\alpha_o) + 0.0305\right)P^4\\&\quad + \left(-0.0019\mathrm{ctg}^2(\alpha_o) - 0.0098\right)P^6\\&\quad + \left(0.0011\mathrm{ctg}^4(\alpha_o) + 0.0015\mathrm{ctg}^2(\alpha_o) - 0.0040\right)P^8\\&\quad + \left(0.0005\mathrm{ctg}^4(\alpha_o) + 0.0001\mathrm{ctg}^2(\alpha_o) + 0.0040\right)P^{10}\ldots\end{aligned} \tag{74}$$

To analyze the characteristics of the electrostatic hexapole-cylindrical energy analyzer under consideration, we have calculated the coefficients of spatial focusing of the first, second, and third orders $\frac{dl}{d\alpha}, \frac{d^2l}{d\alpha^2}, \frac{d^3l}{d\alpha^3}$. The conditions of second-order angular focusing relative to small deviations of a nonequilibrium trajectory from the orbit's center appear as $\frac{dl}{d\alpha} = \frac{d^2l}{d\alpha^2} = 0$. The specific energy dispersion δ, characterizing the energy resolution at a small width of annular slits, also was calculated.

Table 3 of Appendix 2 summarizes the main characteristics of the electron-optical analyzer, calculated as functions of the reflection parameter P and satisfying the conditions of second-order focusing. Analysis of the data in the table shows that the mirror energy analyzer based on hexapole-cylindrical fields allows implementation of the mode of second-order angular focusing when the source and its image are located within the inner cylindrical electrode. The interval of changing the input angles, calculated from the condition of second-order focusing, is small—within the range of 30°–36°. The increase in P and α_0 leads to the increase in the linear energy dispersion D and growth in the cubic angular aberration.

The third-order aberration coefficients are small within the interval $P = 0.2 - 0.8$, while the absolute value of the specific dispersion increases in this range of P values up to a certain maximum value and then starts to decrease, and this takes place most dramatically in the region of $P > 0.8$. Therefore, further calculations in the region of larger values of P seem to have no practical meaning. On the other hand, the increase in the reflection parameter P increases the radial component ρ_m, which in turn requires more accurate calculations, including a series expansion up to the terms having the P^{15} order.

The variant of focusing with the parameters $\alpha_0 = 35.4686°$, $P = 0.7897$, $\Delta = 0.70$, $\rho_m = 0.3050$, $l = 2.7088$, $\delta = -690.30$ was selected as the most optimal energy analyzer scheme (Figure 7). For this scheme, the functions $\Delta l = l(\alpha) - l(\alpha_0)$, characterizing the aberrational broadening of the image at the mirror focus, were calculated. It was established that, for a charged particle beam having an initial angular spread of 8° ($\Delta\alpha = \pm 4$), the image width is small and constitutes 0.2% of the mirror's focal length. This implies that the mirror energy analyzer based on the hexapole cylindrical field can operate in the mode of high resolution and maximal luminosity, which is ensured by the property of spatial focusing in an axisymmetric mirror, and also by the opportunity to focus wide electron beams in the axial plane.

To confirm the validity of the approximate-analytical method in describing the charged particle trajectories, numerical simulation of the electrostatic mirror-type analyzer in question was conducted. Similar to the processes discussed in sections 3 and 4, the calculations were performed using the FOCUS software designated for the simulation of axisymmetric corpuscular-optical systems with a virtually arbitrary geometry of electrodes (Trubitsyn, 2008). Figure 8 shows a plot of charged particle trajectories in the electron-optical system under consideration.

Particles are emitted from a point source positioned on the symmetry axis in the range of polar angles from 30° to 38° and directed onto the detector. As is seen from Figure 8, the secondary electrons (labeled 3 in the figure), excited from the sample under investigation (4) by primary radiation (5), enter the working region through the special window (6) in the inner cylinder covered by a 1D grid. Then secondary electrons are deflected by the field toward the cylinder axis and focused on the cylinder electrode's surface. Those electrons that have passed through the output window's grid (7) in the inner cylinder (1) are registered by the detector (8). The scheme provides the mode of second-order angular focusing of the axis–ring type.

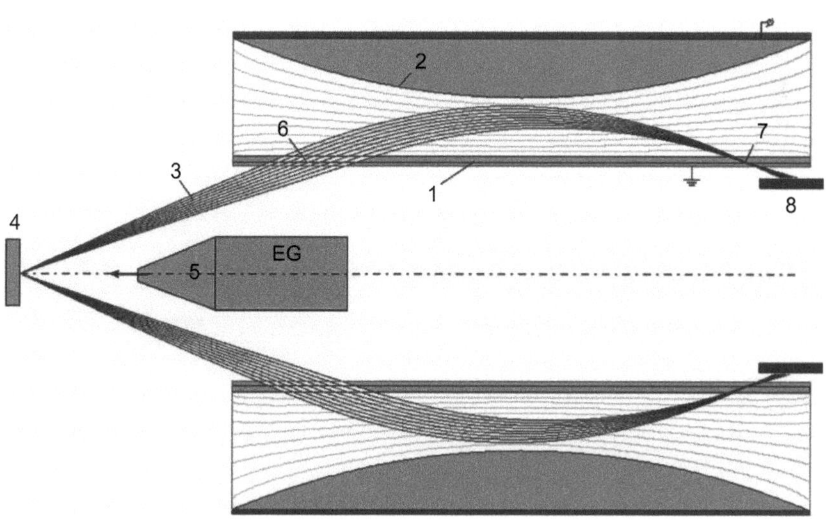

Figure 8 Scheme of the hexapole-cylindrical energy analyzer in the mode of angular focusing axis-ring (1—cylindrical electrode, 2—external concave electrode, 3—secondary electrons, 4—test sample, 5—primary electrons, 6 and 7—input and output slits, 8—detector, EG—electron gun). (See the color plate.)

CHAPTER FOUR

Numerical Methods in the Design of Energy Analyzers

Contents

The main advantage of analytical methods is that they allow establishing relatively simple relationships between the parameters, which is extremely important for revealing general physical regularities inherent to a particular system under study. However, analytical methods assume a high-degree

Advances in Imaging and Electron Physics, Volume 192
ISSN 1076-5670
http://dx.doi.org/10.1016/bs.aiep.2015.08.004

idealization of such a system. Since the conditions of modern experimentation require extremely accurate estimates of output parameters of a device, analytical methods may make a considerable contribution only on the initial, ideological stage of a design, while the responsibility for final engineering solutions, with a variety of features that cannot be described analytically, undoubtedly rests with numerical calculations. The validity of this statement is especially strong with problems of optimization and synthesis, when numerical methods play a decisive role due to the versatility and accuracy of the estimates they grant to researchers.

Numerical modeling (analysis) of electron- and ion-optical systems traditionally includes three sequential stages. These are calculating the electromagnetic fields that govern the charged particle motion, calculating the charged particle trajectories in the system's working area, and finally, calculating the output integral characteristics (functionals) of a device, such as instrumental function, resolution power, and various types of dispersions caused by the presence of aberrations.

1. METHODS FOR NUMERICAL SIMULATION OF ELECTROSTATIC FIELDS

Calculating the potential distribution in an electron/ion-optical system consists of solving the Dirichlet problem for the Laplace equation (or the Poisson equation if the space charge is taken into account) in a certain domain, with the known values of potentials defined on the boundary electrodes. The general mathematical formulation of the problem is as described next.

It is necessary to solve the boundary-value problem for the Poisson equation in the domain Ω with the boundary Γ:

$$\Delta u(\xi) = \rho,\ \xi \in \Omega \cup \Gamma \tag{1}$$

with the boundary condition

$$u(\xi) = \bar{u}(\chi), \chi \in \Gamma, \tag{2}$$

where $\bar{u}(\chi)$ is a piecewise continuous function and ρ is the space charge density.

Three methods are currently popular as basic approaches to the numerical solution of this problem: the finite difference method (FDM), the finite element method (FEM), and the boundary element method (BEM).

FDM (see, for example, Morton & Mayers, 2005) can be most efficiently employed to calculate the fields in the systems with comparatively simple configurations of electrodes, when the nodes of partition mesh fit well the boundary of the calculation domain. The external problem requires introducing a closing surface with artificially imposed boundary conditions—for example, the condition of vanishing of the normal derivative of potential (Manura & Dahl, 2006). Among the finite-difference methods, the estimations of the convergence rate give preference to iterative methods, such as the longitudinal-transverse sweep method with optimal sequence of parameters and the alternating triangular method with Chebyshev's acceleration (Samarskiy, 1983).

FEM (Zienkiewicz, Taylor, & Zhu, 2005) can be successfully used to study systems with a complicated geometry of electrodes. However, irregular arrangement of the mesh nodes for potential calculation complicates the trajectory analysis, which limits the FEM application mainly due to internal problems.

BEM (see, for example, Wrobel & Aliabadi, 2002) represents a genuine numerical realization of the well-known method of integral equations (Cheng & Cheng, 2005) and may be considered the most advanced of the numerical methods of field calculation that have appeared in the computer age. The main advantages of BEM is its simplicity of algorithmization, high accuracy, ability to solve both external and internal problems of potential theory with virtually arbitrary boundary geometry and the presence of different-scale geometry elements.

2. MAIN APPROACHES TO SOLVING THE PROBLEMS OF POTENTIAL THEORY USING BEM

One of the approaches to obtaining an integral relation that delivers a solution to the Dirichlet problem (Brebbia, Telles, & Wrobel, 1984) is to assume that the unknown potential distribution $u(\xi)$ can be expressed in terms of the single-layer potential with the density of sources σ:

$$u(\xi) = \int_{\Gamma} \sigma(\chi) U^*(\xi,\chi) d\Gamma(\chi),\ \xi \in \Omega \cup \Gamma, \tag{3}$$

where the kernel $U^*(\xi,\chi)$ represents either a Newtonian potential or a double-layer potential with density μ:

$$u(\xi) = \int_{\Gamma} \mu(\chi) \frac{\partial U^*(\xi,\chi)}{\partial n} d\Gamma(\chi). \quad (4)$$

Another alternative approach (Brebbia et al., 1984) in electrostatics is based on the second Green formula (Greenberg, 1971):

$$\int_{\Omega} \left(u\nabla^2 w - w\nabla^2 u \right) d\Omega = \oint_{\Gamma} \left(u\frac{\partial w}{\partial n} - w\frac{\partial u}{\partial n} \right) d\Gamma, \quad (5)$$

where n is the outward normal on Γ. In this case, the weight function w is selected as a fundamental solution w^* of the Laplace equation:

$$\nabla^2 w^* = -\delta_\xi 2\alpha\pi, \quad (6)$$

where δ_ξ is the Dirakh delta function concentrated at the point ξ of the domain Ω, α is the coefficient defined by the problem's dimensionality—that is, $\alpha = 1$ and $\alpha = 2$ for two-dimensional (2D) and three-dimensional (3D) cases, respectively. In this case, assuming that the space charge can be neglected ($\rho = 0$), we can represent Eq. (5) as

$$2\alpha\pi u(\xi) + \int_{\Gamma} u(\chi) Q^*(\xi,\chi) d\Gamma(\chi) = \int_{\Gamma} q(\chi) U^*(\xi,\chi) d\Gamma(\chi), \quad (7)$$

with $Q^*(\xi,\chi) = \partial w^*/\partial n$, $q(\chi) = \partial u/\partial n$, $U^*(\xi,\chi) = w^*$, and $\xi \in \Omega$.

The function u here must be a quadratically integrable function.

Eq. (7) implies that the harmonic function $u(\xi)$ can be represented as a sum of the single-layer potential with the density $\sigma = q/2\alpha\pi$ and the double-layer potential with the density $\mu = -u/2\alpha\pi$.

The advantage of Eq. (7) compared to the integral equation [Eq. (3)] for the single-layer potential (Greenfield & Monastyrskiy, 2009) is that the smoothness requirements of the boundary Γ are weakened in the sense that special measures for eliminating the singularities of the density σ on the ribs and vertices of the electrode surfaces are not required (Brebbia et al., 1984). The disadvantage of this alternative approach is that Eq. (7) contains an additional boundary integral compared to Eq. (3).

If $\xi \in \Gamma$, the integral in the left side of Eq. (4) has a jump of discontinuity caused by singularity in the normal derivative $Q^*(\xi,\chi) = \partial w^*/\partial n$. The jump value is determined by the local topological properties of the boundary Γ at the point ξ; i.e., it represents a function of ξ. Thus, we obtain the boundary integral equation

$$\gamma(\xi)u(\xi)+\int_{\Gamma}u(\chi)Q^{*}(\xi,\chi)d\Gamma(\chi)=\int_{\Gamma}q(\chi)U^{*}(\xi,\chi)d\Gamma(\chi), \tag{8}$$

where $\xi \in \Omega \cup \Gamma$. This equation is the starting point for BEM in the direct statement, in contrast to the indirect statement that requires introducing formal single-layer and double-layer densities having no direct relevance to the mathematical statement of the Dirichlet problem (Brebbia et al., 1984).

In the first stage of solving the Dirichlet problem (the inverse problem), an unknown function $q(\chi)$, $\xi \epsilon \Gamma$ has to be found. The second stage (the direct problem) assumes calculation of the potential distribution $u(\xi)$ in the domain of interest $\xi \in \Omega$ using the function $q(\chi)$ previously found. However, along with the undoubted advantages of this method, its serious drawback is the loss of computational accuracy (the computational errors may exceed 100%) in calculating the potential near the boundary, caused by the singular behavior of the functions $Q^{*}(\xi,\chi)$ and $U^{*}(\xi,\chi)$ at $\xi \to \chi$.

3. GENERAL TECHNIQUES FOR CALCULATING THE INTEGRALS OF FUNCTIONS WITH SINGULARITIES

In some cases, it is advisable to attribute, in a conditional sense, a finite numeric value to divergent integrals. A.-L. Cauchy proposed to perform this in the following way (see, for example, Kanwal, 1996). Suppose that the integral

$$\int_{a}^{b} f(x)\,dx \tag{9}$$

has a singularity at the point $x=c$ located in the interior of the interval of integration. Then this singularity is cut out symmetrically and the integral value is defined by passing to the limit as

$$\int_{a}^{b} f(x)\,dx = \lim_{\varepsilon\to+0}\left[\int_{a}^{c-\varepsilon} f(x)\,dx + \int_{c+\varepsilon}^{b} f(x)\,dx\right]. \tag{10}$$

Such a limit may exist even in the case when the integral [Eq. (9)] diverges in a common sense. The limit in the right side of Eq. (10) is called the *principal value* of the integral [Eq. (9)], while the integral itself is called the *singular integral,* in contrast to the proper or convergent improper integrals, which are called *regular integrals.*

Some general techniques are known, which can be useful in the numerical integration of the functions with singularities (Bachvalov, Zhidkov, & Kobel'kov, 1987).

1. ***Multiplicative isolation of singularities.*** The integrand of the integral [Eq. (9)] is represented in the form $f(x)=g(x)p(x)$, where $p(x)$ is a simple function containing a singuliarity, while $g(x)$ is a smooth function. Next, we use the general Gauss approach to the calculation of integrals with the weight p, where the integrand can be approximated by a polynomial function of the order required. This leads to the factorized representation

$$\int_0^1 f(\xi)d\xi = \int_0^1 g(\xi)p(\xi)d\xi = \sum_{i=1}^{N} g(\xi_i)w_i, \tag{11}$$

where ξ_i and w_i are the nodes and weights of the previous quadrature.

When using the BEM, the logarithmic singularity with $p(\xi)=\ln(\xi)$ appears frequently, especially in 2D problems. Detailed tables of nodes and weights for this type of quadrature are commonly used in relevant calculations (Abramovitz & Stegun, 1964). For example, the nodes and weights for $N=4$ are given in Table 1.

2. ***Additive isolation of singularities.*** In some cases, the integrand can be represented as a sum of the singular $\varphi(x)$ and regular $g(x)$ parts: $f(x)=\varphi(x)+g(x)$; herewith the integral $\int_a^b \varphi(x)dx$ can be calculated explicitly, while $g(x)$ is a smooth function that can easily be integrated numerically.

3. ***Change of variable in integration.*** The change of variable $\xi=\varphi(t)$, $\varphi(0)=0$, $\varphi(1)=1$ transforms the original integral $\int_0^1 f(\xi)d\xi$ to $\int_0^1 g(t)dt$,

Table 1 Nodes and Weights of Gaussian Quadratures

ξ_i	w_i
0.041448	0.383464
0.245275	0.386875
0.556165	0.190435
0.848982	0.039225

where $g(t) = f(\varphi(t))\varphi'(t)$. Owing to the presence of the multiplier $\varphi'(t)$, the integrand's singularity can be eliminated.

4. USING BEM TO SOLVE THE INTERIOR DIRICHLET PROBLEM

The interior Dirichlet problem consists of finding a potential distribution in the domain bounded by a surface upon which the boundary conditions of the first kind are defined.

4.1 Numerical Realization

In 3D cases, the boundary Γ is generally partitioned into N elements Γ_j, so the Dirichlet problem [Eq. (8)] takes the discretized form

$$\gamma u(\xi) + \sum_{j=1}^{N} \int_{\Gamma_j} u(\chi) Q^*(\xi, \chi) d\Gamma(\chi) = \sum_{j=1}^{N} \int_{\Gamma_j} q(\chi) U^*(\xi, \chi) d\Gamma(\chi). \tag{12}$$

In the axisymmetric case, this representation entails

$$\gamma u(\xi) + \sum_{j=1}^{N} \int_{\Gamma_j} u(\chi) Q^*(\xi, \chi) R(\chi) d\Gamma(\chi) = \sum_{j=1}^{N} \int_{\Gamma_j} q(\chi) U^*(\xi, \chi) R(\chi) d\Gamma(\chi), \tag{13}$$

where $R(\chi)$ is the value of the coordinate r at the point $\chi \in \Gamma$. The fundamental solution in this case can be expressed through the complete elliptic integral $K(m)$ of type I (Abramovitz & Stegun, 1964):

$$U^*(\xi, \chi) = \frac{4K(m)}{\sqrt{a^* + b^*}}, \tag{14a}$$

$$m = 2b^*/(a^* + b^*), \tag{14b}$$

$$a^* = R^2(\xi) + R^2(\chi) + [Z(\xi) - Z(\chi)]^2, \tag{14c}$$

$$b^* = 2R(\xi)R(\chi). \tag{14d}$$

Here, $Z(\chi)$ is the z-coordinate of the point $\chi \in \Gamma$.

The derivative of the fundamental solution relative to the outward normal to the boundary Γ appears as

$$
\begin{aligned}
Q^*(\xi,\chi) &= A(\xi,\chi)n_R(\chi) + B(\xi,\chi)n_Z(\chi),\\
A(\xi,\chi) &= \frac{4}{\sqrt{a^*+b^*}}\left(\frac{1}{2R(\chi)}\left(\frac{R^2(\xi)-R^2(\chi)+[Z(\xi)-Z(\chi)]^2}{a^*-b^*}\right)E(m)-K(m)\right),\\
B(\xi,\chi) &= \frac{Z(\xi)-Z(\chi)}{a^*-b^*}E(m),
\end{aligned}
\tag{15}
$$

where $E(m)$ is the complete elliptic integral of the II type, and n_Z and n_R are the projections of the unit normal onto axes Z and R, respectively.

Since on each boundary element Γ_j, with the origin at the point χ_j and the end point at χ_{j+1}, the potential u coincides with the potential of the electrode containing the section Γ_j: $u(\chi)=u_j=const, \chi \in \left[\chi_j, \chi_{j+1}\right]$, the following relation holds true:

$$
\sum_{j=1}^{N}\int_{\Gamma_j} u(\chi)Q^*(\xi,\chi)R(\chi)d\Gamma(\chi) = \sum_{j=1}^{N} u_j \int_{\chi_j}^{\chi_{j+1}} Q^*(\xi,\chi)R(\chi)d\Gamma(\chi). \tag{16}
$$

The normal derivative of $q(\chi)$ is more complicated than the normal derivative of $u(\chi)$; therefore, it is commonly interpolated when solving the integral equation [Eq. (13)] numerically. In the simplest case of the piecewise-constant interpolation, a node τ_j is defined at the midpoint of each boundary element Γ_j and the function $q(\chi)$ is assumed to be constant on each Γ_j: $q(\chi)=q(\tau_j)=q_j, \chi \in \left[\chi_j, \chi_{j+1}\right]$.

With regard to this data, Eq. (13) reads as

$$
\gamma u(\xi) + \sum_{j=1}^{N} u_j \int_{\chi_j}^{\chi_{j+1}} Q^*(\xi,\chi)R(\chi)d\Gamma(\chi) = \sum_{j=1}^{N} q_j \int_{\chi_j}^{\chi_{j+1}} U^*(\xi,\chi)R(\chi)d\Gamma(\chi); \tag{17}
$$

The change of variable $x=(\chi-\chi_j)/\Delta\chi_j$ transforms this equation to

$$
\begin{aligned}
&\gamma u(\xi) + \sum_{j=1}^{N} u_j \Delta\chi_j \int_0^1 Q^*\left(\xi, \Delta\chi_j x + \chi_j\right) R\left(\Delta\chi_j x + \chi_j\right) dx\\
&= \sum_{j=1}^{N} q_j \Delta\chi_j \int_0^1 U^*\left(\xi, \Delta\chi_j x + \chi_j\right) R\left(\Delta\chi_j x + \chi_j\right) dx.
\end{aligned}
\tag{18}
$$

Finally, denoting

$$H_j(\xi) = \Delta\chi_j \int_0^1 Q^*\left(\xi, \Delta\chi_j x + \chi_j\right) R\left(\Delta\chi_j x + \chi_j\right) dx, \tag{19a}$$

$$F_j(\xi) = \Delta\chi_j \int_0^1 U^*\left(\xi, \Delta\chi_j x + \chi_j\right) R\left(\Delta\chi_j x + \chi_j\right) dx, \tag{19b}$$

we can represent Eq. (18) in the form

$$\gamma u(\xi) + \sum_{j=1}^{N} u_j H_j(\xi) = \sum_{j=1}^{N} q_j F_j(\xi), \tag{20}$$

where $\gamma = 4\pi$ for $\xi \in \Omega$ and $\gamma = 2\pi$ for $\xi \in \Gamma$ with regard to the jump of the integral $H_j(\xi)$ on the boundary Γ in the case when Γ is approximated by a set of straight segments Γ_j.

Electrostatic field calculation is performed in two stages. First, using Eq. (20), the unknown vector q_j is determined in accordance with the boundary potential distribution ($\xi \in \Gamma$); i.e., the inverse problem of defining the cause according to its consequence is solved. Then the found values q_j and the given values u_j are used to determine the function $u(\xi)$, $\xi \in \Omega$ from Eq. (20), or the direct problem of finding the consequence generated by its cause is solved.

In general, two factors impair the accuracy of solving Eq. (20): the degree of partition of the boundary Γ (or, in other words, the value of N,) and the accuracy of calculating the integrals $H_j(\xi)$, $F_j(\xi)$. If the integrands are regular functions, the conventional Gauss quadratures with a moderate number of nodes ensure a sufficient accuracy of this integration. However, the logarithmic singularity that appears in the integrands when $\xi \to \Gamma_j$ may lead to considerable accuracy losses.

Next, we propose a technique of weakening the singularities, provided that the boundary of the domain Ω is composed of a set of rectilinear segments (Trubitsyn, 1995a).

Accuracy estimation for the technique proposed here has been derived using a test domain, the boundary of which consists of inner and external cylinders with the radii of 1 and 2, respectively, each having a length of 10. The domain's geometry also includes two closing rings supplied with the potential of 0.5, placed at both ends of the cylindrical electrodes in

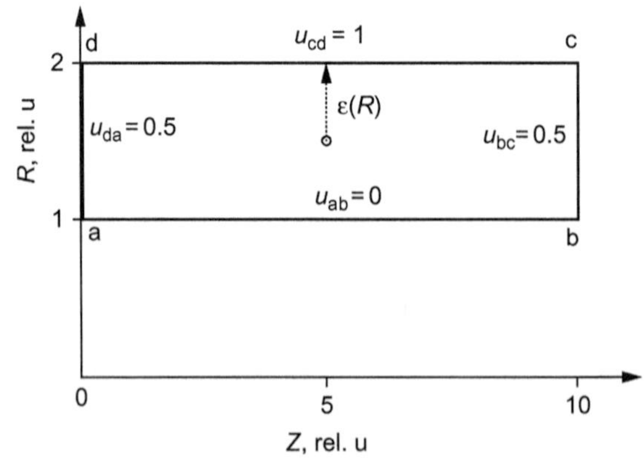

Figure 1 Test domain geometry.

the space between them (Figure 1). The upper half of the transaxial cross section of this design is a rectangle of length 10 and width 1. The reference potential distribution in this region was calculated using the alternating triangular method with Chebyshev acceleration (Samarskiy, 1983).

Radial dependence of the relative error $\varepsilon(R)$ of potential calculation was interpolated at the reference points having a z-coordinate equal to 5 and R-coordinates equal to 1.5, 1.6, 1.7, 1.8, 1.9, 1.95, 1.97, 1.98, and 1.99. When using the BEM, the rectangular boundary was divided into 44 boundary elements of equal length. The integration over each boundary element Γ_j was performed using the Gaussian quadrature formulas with four nodes.

4.2 Calculating Integrals of Functions with Singularities in Solving the Inverse Problem

The inverse problem is solved by means of the collocation method, according to which points ξ_i are defined in the middle of each element Γ_i ($\xi_i = \tau_i$), and a system of N equations is formed for a whole set of the points ξ_i ($i = 1, \ldots, N$); see Eq. (20):

$$2\pi u(\xi_i) + \sum_{j=1}^{N} u_j H_{ij} = \sum_{j=1}^{N} q_j F_{ij}, \quad i = 1, 2, \ldots, N, \tag{21}$$

where $H_{ij} = H_j(\xi_i)$ and $F_{ij} = F_j(\xi_i)$.

The singularity in the integrals takes place at $i = j$; i.e., when the integration is performed over the element Γ_j (we call it the singular element) containing the current collocation node ξ_i.

1. **Calculation of H_{ii}.** Let us define a constant potential u_Γ on the entire boundary Γ. Then the normal derivative on the boundary is zero; i.e., $q_j=0$, $j=1, 2, \ldots, N$. In that case, the latter equation appears as

$$2\pi u_\Gamma + u_\Gamma H_{ii} + u_\Gamma \sum_{j=1,\, i\neq j}^{N} H_{ij} = 0, \tag{22}$$

which entails that

$$2\pi + H_{ii} = -\sum_{j=1,\, i\neq j}^{N} H_{ij}. \tag{23}$$

Thus, the singular integral H_{ii} can be calculated according to Eq. (23) through the sum of the nonsingular integrals (see also Brebbia et al., 1984).

In a particular case, a more accurate estimate can be obtained from the following considerations. Since for singular planar boundary elements in 3D problems, as well as for singular line elements in axisymmetric and 2D problems, we have $\mathrm{grad}\, U^*(\xi,\chi)\perp \mathbf{n}$, then $Q^*(\xi,\chi)=\partial U^*/\partial n = \mathrm{grad}\, U^*(\xi,\chi)\cdot \mathbf{n}=0$, so we obtain

$$H_{ii} = 0 \tag{24}$$

in the interior Dirichlet problem.

Figure 2 (curve 1) shows radial dependence of the potential calculation error in the test problem when calculating H_{ii} by means of the procedure described here. In the case of calculating H_{ii} in accordance with Eq. (19a), the relevant error exceeds 100% at all points of the calculation domain.

2. **Calculation of F_{ii}.** Let us represent the complete elliptic integral of the first kind in the polynomial form (Abramovitz & Stegun, 1964):

$$K(m) = B(m_1)\text{-}A(m_1)\ln m_1, \tag{25}$$

where $m_1=1-m$, $B(m_1)=\sum_{n=0}^{N_K} p_n m_1^n$, $A(m_1)=\sum_{n=0}^{N_K} s_n m_1^n$, p_n, s_n are the tabulated coefficients of the polynomial representation, while the value m is calculated according to Eq. (14b).

The R- and z-coordinates of the points belonging to the interval Γ_i, can be expressed through the relations

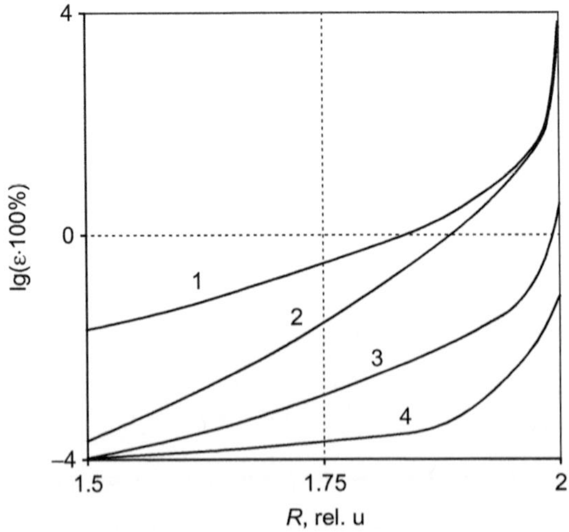

Figure 2 Radial dependence of the potential calculation error.

$$R(\Delta\chi_i x + \chi_i) = R(x) = c_i x + d_i,\ Z(\Delta\chi_i x + \chi_i) = Z(x) \\ = a_i x + b_i, \tag{26}$$

where $c_i = sin(\alpha)\Delta\chi_i$, $a_i = cos(\alpha)\Delta\chi_i$; b_i, d_i are the coordinates of the origin point of the element Γ_i, α is its angle of inclination to the z-axis, $x \epsilon [0, 1]$.

In this case [see Eqs. (14c) and (14d)], $a^* - b^* = (\xi_i - x)\ ^2\Delta\chi_i^{\ 2}$, $a^* + b^* = [c_i\ (\xi_i + x) + 2d_i] + a_i^2\ (\xi_i - x)^2$. Then

$$\ln m_1 = \ln\frac{a^* - b^*}{a^* + b^*} = \ln(\xi_i - x)^2 - \ln\frac{a^* + b^*}{\Delta\chi_i^2} \\ = 2\ln|\xi_i - x| - \ln\frac{a^* + b^*}{\Delta\chi_i^2}, \tag{27}$$

and, in turn,

$$K(m) = B(m_1) + A(m_1)\ln\frac{a^* + b^*}{\Delta\chi^2} - 2A(m_1)\ln|\xi_i - x|, \tag{28}$$

$$B(m_1) = \sum_{n=0}^{N_k} p_n m_1^n = \sum_{n=0}^{N_k} p_n \left[\frac{(\xi_i - x)^2\Delta\chi_i^2}{a^* + b^*}\right]^n, \tag{29}$$

$$A(m_1) = \sum_{n=0}^{N_k} s_n m_1^n = \sum_{n=0}^{N_k} s_n \left[\frac{(\xi_i - x)^2\Delta\chi_i^2}{a^* + b^*}\right]^n. \tag{30}$$

Thus, the function

$$U^*(\xi_i, x) = 4G_0(\xi_i, x)/R_i(x) - \frac{8A(m_1)}{\sqrt{a^* + b^*}}\ln|\xi_i - x| \tag{31}$$

has an explicitly logarithmic order of singularity. To separate this singularity, let us represent the original integral [Eq. (19b)] via the difference

$$F_{ii} = F_{1ii} - F_{2ii}, \tag{32}$$

where

$$F_{1ii} = 4\Delta\chi_i \int_0^1 G_0(\xi_i, x)dx, F_{2ii} = 8\Delta\chi_i \int_0^1 \frac{A(m_1)}{\sqrt{a^* + b^*}}\ln|\xi_i - x|R_i(x)dx. \tag{33}$$

Inasmuch as ξ_i is located at the center of the ith interval (i.e., $\xi_i = 1/2$), we can write the following expression for F_{2ii}:

$$F_{2ii} = 8\Delta\chi_i \left[\int_0^{1/2} \frac{A(m_1)}{\sqrt{a^* + b^*}} \ln(1/2 - x)R_i(x)dx + \int_{1/2}^1 \frac{A(m_1)}{\sqrt{a^* + b^*}} \ln(x - 1/2)R_i(x)dx \right]. \tag{34}$$

Let us perform the following transformations of variables. In the first integral, we put $t = 1 - 2x$. Then $a^* + b^* = [c_i\ (1 - t/2) + 2d_i]^2 + a_i^2 t^2/4$, $R_i\ (t) = [(1 - t)c_i/2 + d_i]$, $dt = -2dx$, so that the lower and upper integration limits becomes 1 and 0, respectively. In the second integral, we put $t = 2x - 1$. Then $a^* + b^* = [c_i\ (1 + t/2) + 2d_i]^2 + a_i^2 t^2/4$, $R_i\ (t) = [(1 + t)c_i/2 + d_i]$, $dt = 2dx$, so that the lower and upper integration limits becomes 0 and 1, respectively.

From this, we obtain

$$F_{2ii} = 4\Delta\chi_i \left\{ \int_0^1 (G_i(-t) + G_i(t))\ln(t)dt - \ln 2 \int_0^1 (G_i(-t) + G_i(t))dt \right\}, \tag{35}$$

where

$$G_i(t) = ((1+t)c_i + 2d_i)\frac{\sum_{n=0}^{N_k} s_n \left(\frac{(t\Delta\chi)^2}{[c_i(2+t)+4d_i]^2 + (ta_i)^2}\right)^n}{\sqrt{[c_i(2+t)+4d_i]^2 + (ta_i)^2}}. \tag{36}$$

Finally, renaming t as x, we have

$$F_{ii} = 4\Delta\chi_i \left\{ \int_0^1 [G_0(\xi_i, x) + \ln 2(G_i(-x) + G_i(x))]dx + \int_0^1 (G_i(-x) + G_i(x))\ln(1/x)dx \right\}. \tag{37}$$

The first integral in Eq. (37) can be calculated using the conventional Gauss quadrature, while numerical integration of the functions having the form $f(x)\ln(1/x)$ can be performed using a special quadrature with the weight $\ln(1/x)$ (see section 3) that ensures the calculation accuracy required. The multiplicative method of singularity separation based on Eq. (37) has a number of advantages over the additive method that is most often used for these purposes (Freykman, 1980; Greenfield & Monastyrskiy, 2009). In particular, it provides complete control over the absolute error by varying the number of summands in the relevant quadrature formulas. It should be added that, even in the case of a small number of relevant coefficients ($N_k = 5 \div 7$), the errors in calculating the complete elliptic integral via the polynomial representation [Eq. (25)] remain less compared to the roundoff errors in the 8-byte data representation.

Thus, in accordance with Eq. (21), a solution q_j of the inverse problem can be obtained from the following system of linear equations:

$$b_i = F_{ij}q_j, \tag{38}$$

where $b_i = 2\pi u_i + \sum_{j=1}^{N} u_j H_{ij}$, $i = 1, 2, ..., N$; herewith, the integrals H_{ij} and F_{ij} for $i \neq j$ are calculated according to Eqs. (19a) and (19b), and for $i = j$ according to Eqs. (24) and (37), respectively.

Curve 2 in Figure 2 illustrates an improvement in the accuracy of potential calculation in the test problem, when the singular integrals H_{ii} and F_{ii} are calculated using the techniques described.

4.3 Calculation of Quasi-singular Integrals in Solving the Direct Problem

The direct problem solution is given by the formula [see Eq. (20)]:

$$u(\xi) = \frac{1}{4\pi}\left[\sum_{j=1}^{N} u_j H_j(\xi) - \sum_{j=1}^{N} q_j F_j(\xi)\right], \ \xi \in \Omega. \tag{39}$$

Curves 1 and 2 in Figure 3 show that the closer the point ξ is to the boundary element Γ_j, the stronger is the manifestation of the extremum in the integrand [Eq. (19b)] (in the limiting case, a discontinuity arises), which considerably complicates direct application of the Gauss quadrature.

Next, we demonstrate the possibility for improving the calculation accuracy of the quasi-singular integrals $H_j(\xi)$ and $F_j(\xi)$, which in turn significantly reduces the error in calculating $u(\xi)$.

1. **Calculating $H_j(\xi)$.** Let us define a constant potential u_Γ on the boundary Γ. Again, we may put $q_j = 0, j = 1, 2, \ldots, N$. In accordance with the maximum principle $u(\xi) = u_\Gamma$. In this case, the following equation holds true for any $\xi \in \Omega$ [see Eq. (12)]:

$$4\pi u(\xi) + u_\Gamma \sum_{j=1}^{N} H_j(\xi) = 0, \tag{40}$$

where $H_j(\xi)$ is the exact integral value. Since $u(\xi) = u_\Gamma$, then

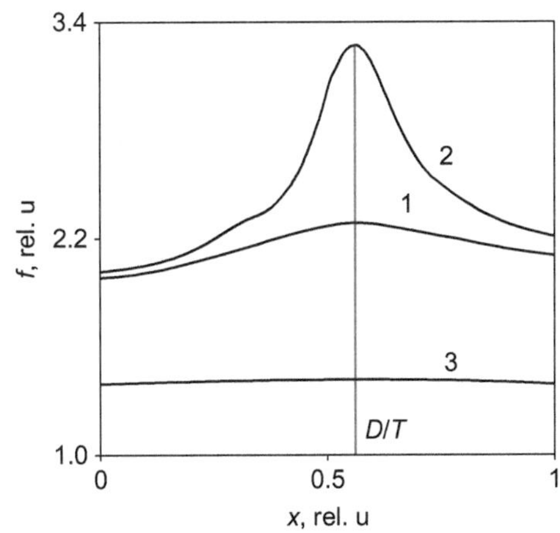

Figure 3 Integrands: 1—$R(\xi) = 0.9$; 2—$R(\xi) = 0.98$; 3—with separate singularities.

$$4\pi + \sum_{j=1}^{N} H_j(\xi) = 0. \tag{41}$$

Numerical values of the integrals $H_j^*(\xi)$, calculated according to Eq. (19a), have a certain error ε_j, so that $H_j(\xi) = H_j^*(\xi) - \varepsilon_j$. The greater the magnitude of the error ε_j, the stronger is the integrand singularity, which, in turn, increases with decreasing the distance between the boundary element Γ_j and the point ξ. A priori information about the nature of the error ε_j allows us to conclude that ε_j is zero when this distance is infinitely large and becomes infinite at zero distance, which entails that

$$\varepsilon_j = P(\xi)/(r_j)^k, \tag{42}$$

where r_j is the distance between the point ξ and the closest point to the segment Γ_j, $P(\xi)$ is unknown parameter, k is the exponent that should be determined empirically.

Thus, we can write that

$$H_j(\xi) = H_j^*(\xi) - P(\xi)/(r_j)^k, \tag{43}$$

which allows determination of the unknown parameter $P(\xi)$ from Eq. (41):

$$P(\xi) = \frac{\sum_{j=1}^{N} H_j^* + 4\pi}{\sum_{j=1}^{N} (r_j)^{-k}}. \tag{44}$$

The simplest procedure for calculating the integrals $H_j(\xi)$ in accordance with Eqs. (44) and (43) improves the accuracy of potential calculation near the boundary by more than two orders of magnitude, so that the calculation error turn out less than 3% for the test problem (see curve 3 in Figure 2). The highest accuracy is achieved for $k=4$.

2. **Calculating $F_j(\xi)$.** It is required to calculate the integral of type

$$I = \int_0^1 g(x) f(x)\, dx, \tag{45}$$

where $g(x)$ belongs to the class of infinitely differentiable functions and the function $f(x)$ infinitely increases at the point $x = a \in [0,1]$.

To solve this problem, we use the formalism of the well-known additive method (Greenfield & Monastyrskiy, 2009), the essence of which lies in the representation of the function with a singularity in the form of a sum of two functions. The first of these functions contains the singularity and herewith allows analytical integration, while the second does not contain any singularities and is integrated numerically. To this end, we represent the integral I as follows:

$$I=\int_0^1 [g(x)-g(a)]f(x)dx+g(a)\int_0^1 f(x)dx=\int_0^1 [g(x)-g(a)]f(x)dx+g(a)J, \tag{46}$$

where the integral $J=\int_0^1 f(x)dx$ can be calculated analytically.

Let us study the integral $\int_0^1 [g(x)-g(a)]f(x)dx$. We expand $g(x)$ into the Taylor series in the vicinity of $x=a$: $g(x)=g(a)+g'(a)(x-a)+\ldots$. Then

$$\int_0^1 [g(x)-g(a)]f(x)dx=\int_0^1 [g'(a)(x-a)+\ldots]f(x)dx. \tag{47}$$

If the function $f(x)$ satisfies the condition $\lim_{x\to a}(x-a)f(x)=0$, the function$[g(x)-g(a)]f(x)$ is infinitely small at $x=a$ and is bounded on the segment [0,1], so that the corresponding integral can be calculated with the desired accuracy using the Gauss quadrature.

We have the integral

$$F_j(\xi)=\Delta\chi_j\int_0^1 U^*(\xi,x)R(x)dx, \tag{48}$$

where, according to Eq. (14a), the function $U^*(\xi,x)$ is expressed in terms of the elliptic integral, which in turn can be calculated from Eq. (25). With regard to this data, the integral $F_j(\xi)$ appears as

$$F_j(\xi)=4\Delta\chi_j\left[\int_0^1 \frac{B(m_1)+A(m_1)\ln(a^*+b^*)}{\sqrt{a^*+b^*}} R(x)dx-\int_0^1 \frac{A(m_1)\ln(a^*-b^*)}{\sqrt{a^*+b^*}}R(x)dx\right]. \tag{49}$$

The integrand in the first integral has no singularities. Taking into consideration Eqs. (14c) and (14d), the second integral can be written in the form

$$\int_0^1 g(x)\ln(a^* - b^*)dx = \int_0^1 g(x)\ln\left(Tx^2 - 2Dx + P\right)dx, \tag{50}$$

where $g(x) = A(m_1)R(x)/\sqrt{a^* + b^*}$, $T = \Delta\chi_j$, $D = [R(\xi)\text{-}d_j]c_j + [Z(\xi)\text{-}b_j]a_j$, and $P = [R(\xi)\text{-}d_j]^2 + [Z(\xi)\text{-}b_j]^2$.

A jump in the integrand is observed at low difference $(a^* - b^*) > 0$; i.e., at the point of minimum $Tx^2 - 2Dx + P$. Inasmuch as the condition of minimum is fulfilled at $x = D/T$, we obtain in the frame of the formalism described:

$$\begin{aligned}\int_0^1 g(x)\,\ln(a^* - b^*)dx = \int_0^1 \left[g(x) - g\left(\frac{D}{T}\right)\right]\,\ln\left(Tx^2 - 2Dx + P\right)dx \\ + g(D/T)\int_0^1 \ln\left(Tx^2 - 2Dx + P\right)dx.\end{aligned} \tag{51}$$

The latter integral can be expressed analytically, which gives a formula to calculate $F_j(\xi)$:

$$F_j(\xi) = 4\Delta\chi_j \left\{ \begin{aligned} &\int_0^1 \left[\frac{K(m)}{\sqrt{a^* + b^*}}R(x) + g(D/T)\,\ln\left(Tx^2 - 2Dx + P\right)\right]dx - \\ &-g(D/T)\left[\begin{aligned}&(1 - D/T)\,\ln(T - 2D + P) + D\,\ln(P)/T - 2 + \\ &+2\frac{\sqrt{TP - D^2}}{T}\left(\text{arctg}\frac{D}{\sqrt{TP - D^2}} + \text{arctg}\frac{T - D}{\sqrt{TP - D^2}}\right)\end{aligned}\right]\end{aligned}\right\}. \tag{52}$$

When the point ξ is approaching the element Γ_j, the function $Tx^2 - 2Dx + P$ tends to $T(x - D/T)^2$, which indicates the absence of a singularity in the integrand $\frac{K(m)}{\sqrt{a^* + b^*}}R(x) + g(D/T)\ln\left(Tx^2 - 2Dx + P\right)$ in the interval $[0,1]$.

Figure 3 demonstrates graphically the proposed method of weakening the singularities. Curve 2 represents the function $K(m)R(x)/\sqrt{a^* + b^*}$, which has a singularity within the element Γ_j located at the closest distance from the point ξ, the radial coordinate of which is 1.98. Curve 3 is the function $K(m)R(x)/\sqrt{a^* + b^*} + g(D/T)\,\ln\left(Tx^2 - 2Dx + P\right)$. The comparison of curves 2 and 3 confirms the deduction on the singularity elimination in the original function. The possibility of employing the conventional Gauss quadrature for calculation of integrals when determining $F_j(\xi)$ is illustrated by curve 4 in Figure 2, which demonstrates improvement of the accuracy of potential calculation in the test problem by more than an order of magnitude compared to the previous case.

5. EXTERIOR DIRICHLET PROBLEM AND CALCULATING THE INTEGRALS OF FUNCTIONS WITH SINGULARITIES

The exterior Dirichlet problem consists of finding the potential in the unrestricted domain G with a given potential distribution on the boundary Γ representing a set of closed contours $\Gamma_1, \Gamma_2, \ldots, \Gamma_k$ (Figure 4).

In this case, a fictitious spherical boundary $\overline{\Gamma}$ with the radius $R_0 \rightarrow \infty$, encircling the boundary Γ, is introduced (Figure 5). The boundary integral equation for the domain Ω restricted by the real boundary Γ and the fictitious boundary $\overline{\Gamma}$ can be written as

$$\begin{aligned} \gamma u(\xi) + \int_{\Gamma} u(\chi) Q^*(\xi,\chi) d\Gamma(\chi) + \int_{\overline{\Gamma}} u(\chi) Q^*(\xi,\chi) d\Gamma(\chi) \\ = \int_{\Gamma} q(\chi) U^*(\xi,\chi) d\Gamma(\chi) + \int_{\overline{\Gamma}} q(\chi) U^*(\xi,\chi) d\Gamma(\chi). \end{aligned} \tag{53}$$

In 3D problems, the asymptotics of the function $u(\chi)$ at $R_0 \rightarrow \infty$ is at least $0(R_0^{-1})$, while its derivative has the asymptotics $q(\chi) = 0(R_0^{-2})$. This implies that the boundary integral equation for interior points of the infinite regular domain coincides with the equation for the finite domain; i.e., it has the form of Eq. (8).

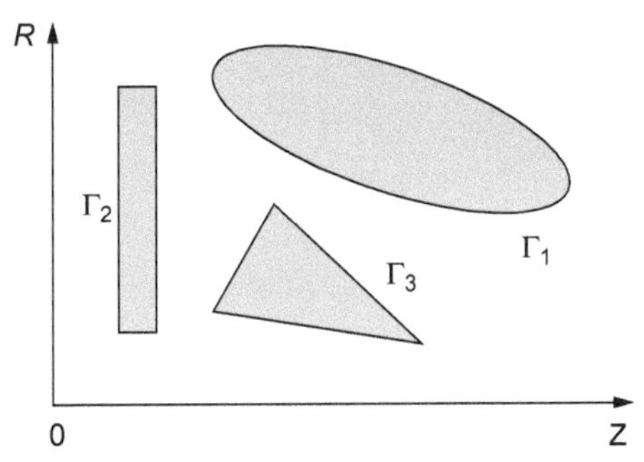

Figure 4 The multiply connected boundary Γ as a set of contours Γ_1, Γ_2, Γ_3.

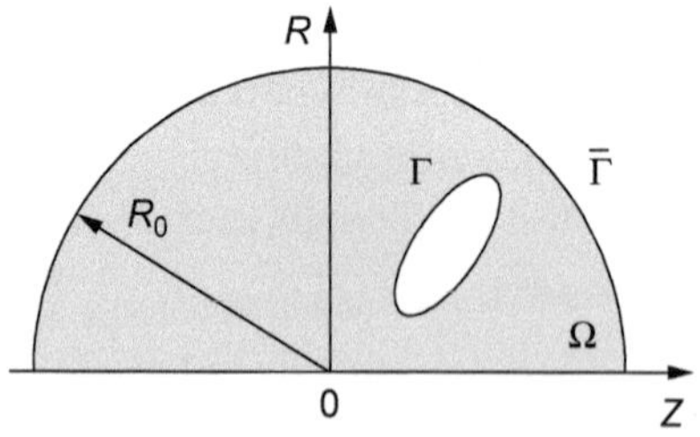

Figure 5 Geometry of the domain with a cavity.

1. **Inverse problem ($\xi \in \Gamma$).**

As we have already noted, for inverse problem we have $\gamma = 2\pi$. Let us define a constant potential u_Γ on the interior and external boundary contours. In that case, the normal derivative, both on the interior (Γ) and the external ($\overline{\Gamma}$) boundaries, equals zero; i.e., $q(\chi) = 0$. This implies that Eq. (53) takes the form

$$2\pi + \int_{\Gamma} Q^*(\xi,\chi)d\Gamma(\chi) + \int_{\overline{\Gamma}} Q^*(\xi,\chi)d\Gamma(\chi) = 0. \tag{54}$$

Since the normal derivative of the fundamental solution on the exterior boundary $\overline{\Gamma}$ $Q^* = \partial U^*/\partial R = \partial(1/R)/\partial R = -1/R^2$ and $d\overline{\Gamma} = R^2 \sin\theta \times d\theta \times d\varphi$, where θ and φ are the azimuthal and polar angles in spherical coordinate system, we have

$$\int_{\overline{\Gamma}} Q^*(\xi,\chi)d\Gamma(\chi) = -4\pi, \tag{55}$$

and, therefore

$$2\pi + \int_{\xi} Q^*(\xi,\chi)d\Gamma(\chi) + \int_{\Gamma-\xi} Q^*(\xi,\chi)d\Gamma(\chi) = 4\pi, \tag{56}$$

where the integral over ξ has a meaning of the integral over an infinitely small part of the boundary, containing the point ξ.

Thus, Eq. (56) allows calculation of the singular integral $\int_\xi Q^*(\xi,\chi)d\Gamma(\chi)$ through the nonsingular $\int_{\Gamma-\xi} Q^*(\xi,\chi)d\Gamma(\chi)$.

As already noted, since $\operatorname{grad} U^*(\xi,\chi) \perp \mathbf{n}$ for singular boundary elements, we have $Q^*(\xi,\chi) = \partial U^*/\partial n = \operatorname{grad} U^*(\xi,\chi) \cdot \mathbf{n} = 0$. This entails that $H_{ii} = 0$ for the interior Dirichlet problem. Due to a jump in double-layer potential when intersecting the boundary from the inside out (Courant & Hilbert, 1951), we come to the following exact value of the integral H_{ii} for the external problem:

$$H_{ii} = 4\pi. \tag{57}$$

The numerical solution q_j of the inverse problem can be obtained from the system of linear equations:

$$b_i = F_{ij} q_j, \tag{58}$$

where $b_i = 2\pi u_i + \sum_{j=1}^{N} u_j H_{ij}$, $i = 1, 2,..., N$; herewith, for $i \neq j$ the integrals H_{ij} and F_{ij} are calculated according to Eqs. (19a) and (19b), while for $i = j$ according to Eqs. (57) and (37).

2. **Direct problem ($\xi \in \Omega$).**

   As we have already established, $\gamma = 4\pi$ for the exterior problem. Assume that the interior closed contours (boundary Γ) and the exterior contour with the radius $R_0 \to \infty$ (boundary $\overline{\Gamma}$) possess a constant potential u_Γ. In that case, the normal derivative of the potential on the boundary $q(\chi) = 0$, and the integral equation [Eq. (53)] for interior points of the 3D infinite region appears as

$$4\pi + \int_{\Gamma} Q^*(\xi, \chi) d\Gamma(\chi) + \int_{\overline{\Gamma}} Q^*(\xi, \chi) d\Gamma(\chi) = 0. \tag{59}$$

The estimate [Eq. (55)] for the second boundary integral gives the equality $\int_{\overline{\Gamma}} Q^*(\xi, \chi) d\Gamma(\chi) = 0$, which implies that for 3D (including axisymmetric) exterior boundary problems,

$$\sum_{j=1}^{N} H_j(\xi) = 0. \tag{60}$$

The assumption [Eq. (42)] about the nature of the error arising in calculating the integrals $H_j^*(\xi)$ in accordance with Eq. (19a) entails a more precise numerical estimate:

$$H_j(\xi) = H_j^*(\xi) - P(\xi) / \left(r_j\right)^k, \tag{61}$$

where the parameter $P(\xi)$ is determined from Eq. (60):

$$P(\xi) = \frac{\sum_{j=1}^{N} H_j^*}{\sum_{j=1}^{N} (r_j)^{-k}}. \quad (62)$$

The quasi-singular integrals $F_j(\xi)$ are calculated according to Eq. (52) obtained when considering the internal problem.

This technique for solving the external Dirichlet problem has been tested by calculating the potential distribution $u(\xi)$ generated in the surrounding space by a sphere with unit potential. The calculation accuracy for the function $u(\xi) = 1/R$ lies within $10^{-4} - 10^{-6}$ % for 100–200 boundary partition elements.

Thus, we may deduce that the proposed technique for evaluating the integrals with singular and quasi-singular integrands allow high-accuracy potential calculation at any point in space for a set of electrodes of virtually arbitrary configuration.

6. NUMERICAL CALCULATION OF POTENTIAL GRADIENT

The right side of the differential equations of motion [see the system of Eq. (71)] is determined by the potential gradient and represents a force function that affects the charged particles. In the planar and axisymmetric problems, the potential depends on two variables: $U = U(x,y)$; therefore $\mathbf{E}(x, y) = -\mathrm{grad}\, U = E_x\boldsymbol{i} + E_y\boldsymbol{j}$, where the gradient components are $E_x = -\partial U/\partial x$, $E_y = -\partial U/\partial y$. If the potential distribution is found numerically (say, using BEM), the partial derivatives in the latter expressions can be calculated using the numerical differentiation formulas (Abramovitz & Stegun, 1964) constructed on the mesh nodes with coordinates (x_i, y_j), where $i = 1,2\ldots n_x$; $j = 1,2\ldots n_y$. Here, n_x and n_y represent the total number of partitions in the x- and y-directions, respectively.

The analysis shows that, despite the existence of effective numerical means for calculating the potential distribution for a wide class of geometries (including FDM, FEM, and BEM), along with trajectory analysis techniques that are well tested in practice, the numerical evaluation of potential gradients calculated on a 2D discrete mesh remains a bottleneck in computational charged particle optics.

The potential distribution undergoes a kink upon the electrode surface (either infinitely thin or having a real thickness), while the field intensity acquires a jump of discontinuity. To investigate the nature of errors in numerical differentiation of potential distributions, next we consider a model potential distribution *U(r)* in a cylindrical capacitor (Figure 6) with the radius R_1 of the inner grounded cylinder 1 and the radius $R_2=\exp(1)$ of the outer cylinder that is charged to the potential $V=1$. The radial coordinate r is changing from 0 to 4. The potential distribution *U(r)* is defined as $U(r)=0$ for $0\leq r\leq R_1$, $U(r)=1$ for $R_2\leq r\leq 4$, and

$$U(r)=V\frac{\ln(r/R_1)}{\ln(R_2/R_1)} \tag{63}$$

for $R_1\leq r\leq R_2$ (Figure 7a).

Radial dependence of the potential gradient's module in the cylindrical capacitor within the interval $0\leq r\leq 4$,

$$E=-\frac{dU}{dr}=-\frac{1}{r}, \tag{64}$$

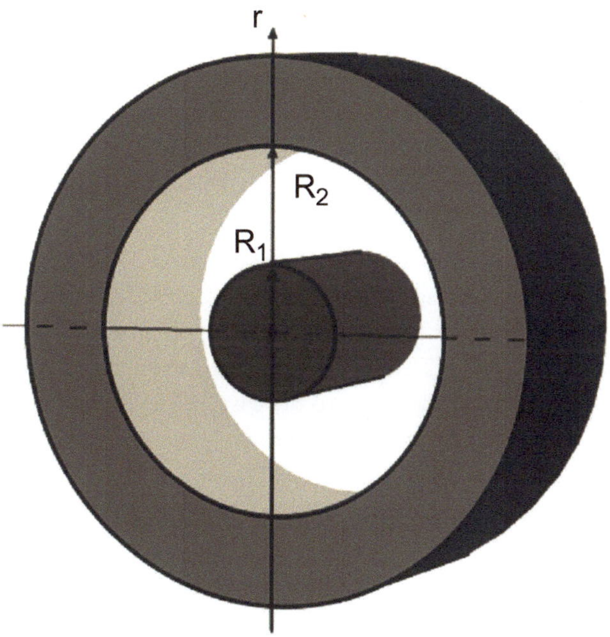

Figure 6 Cylindrical capacitor with real-thickness electrodes. (See the color plate.)

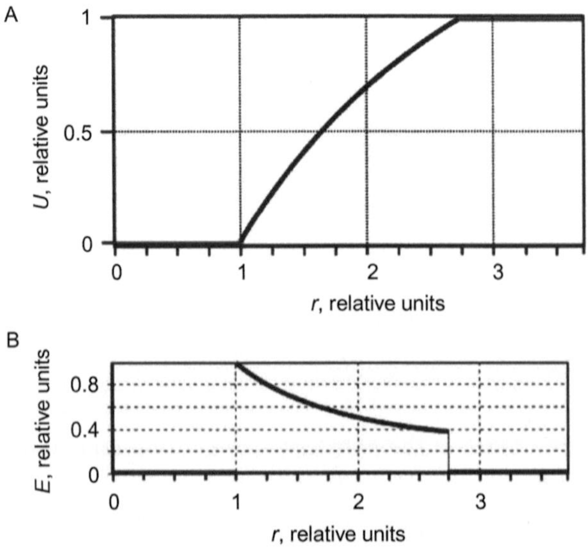

Figure 7 Radial dependence of the potential distribution (A) and potential gradient's module (B) in the cylindrical capacitor with real-thickness electrodes. (See the color plate.)

is shown in Figure 7b. Further, we use this function as a benchmark for comparative analysis of different approaches to numerical differentiation of potential distributions.

6.1 Investigating the Accuracy of Potential Gradient Calculation Using Finite-Difference Formulas

The use of interpolation techniques (for example, see Abramovitz & Stegun (1964) allows for obtaining the numerical differentiation formulas of any accuracy order. Consider a function $U(r)$ defined on a discrete set of $N+1$ nodes ($r_i = ih$, $i = 0,1,2 \ldots N$). By applying the Lagrange interpolation polynomial and its remainder term to the case of five nodes and introducing the notation $U(r_i) = U_i$, we obtain the following commonly used approximations of the derivatives (Abramovitz & Stegun, 1964):

$$E_0 = U_0' = \frac{1}{12h}(-25U_0 + 48U_1 - 36U_2 + 16U_3 - 3U_4) + O(h^4),$$

$$E_1 = U_1' = \frac{1}{12h}(-3U_0 - 10U_1 + 18U_2 - 6U_3 + U_4) - O(h^4),$$

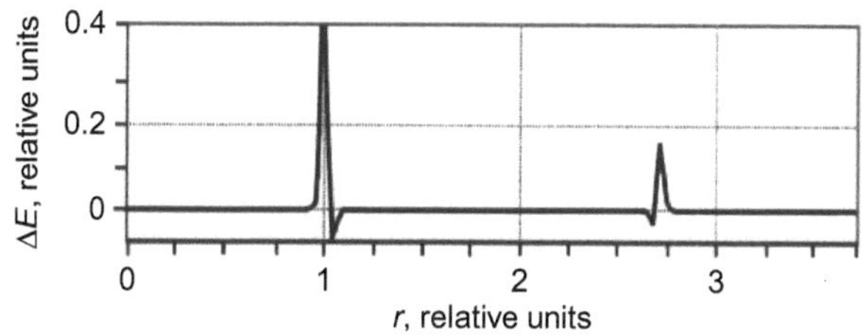

Figure 8 The accuracy of differentiation using the finite-difference formulas. (See the color plate.)

$$E_i = U_i' = \frac{1}{12h}(U_{i-2} - 8U_{i-1} + 8U_{i+1} - U_{i+2}) + O(h^4),\ i = 2, 3, \ldots, N\text{-}2,$$

$$\begin{aligned} E_{N-1} &= U_{N-1}' \\ &= \frac{1}{12h}(-U_{N-4} + 6U_{N-3} - 18U_{N-2} + 10U_{N-1} + 3U_N) - O(h^4), \\ E_N &= U_N' \\ &= \frac{1}{12h}(3U_{N-4} - 16U_{N-3} + 36U_{N-2} - 48U_{N-1} + 25U_N) + O(h^4). \end{aligned} \tag{65}$$

The result of using these formulas for $N = 100$ is shown in Figure 8. The value $\Delta E / E_{max}$ reflects the relative error of derivative calculation, where ΔE is the absolute deviation of the calculated value from the exact analytical value and E_{max} is the exact analytical value at its maximum, which equals 1 in the case under consideration (see Figure 7b). Numerical estimates of the relative errors of the approximations that have been used here are as follows: the mean error is 0.00739, and the maximum error is 0.44255.

As might be expected, the use of standard finite-difference formulas leads to large errors in the vicinity of electrodes ($r = R_1$ and $r = R_2$).

6.2 Investigating the Accuracy of Potential Gradient Calculation Using Shifted Finite-Difference Formulas

We have seen that the standard finite-difference formulas may possess low calculation accuracy. The algorithm we propose here to solve this problem suggests the use of a priori information on the behavior of a differentiable function within different parts of the calculation domain. This gives grounds for choosing the proper finite-difference formulas based on the left, right, or central differences. The relevant formulas, with some necessary notations, are presented next.

Calculation of the derivatives using the left differences:

$$E_i = U_i' = E_i^L = (3U_{i-4} - 16U_{i-3} + 36U_{i-2} - 48U_{i-1} + 25U_i)/12h, \quad (66a)$$

$$E_i = U_i' = E_i^{LC} = (-U_{i-3} + 6U_{i-2} - 18U_{i-1} + 10U_i + 3U_{i+1})/12h. \quad (66b)$$

Calculation of the derivatives using the central differences:

$$E_i = U_i' = E_i^C = (U_{i-2} - 8U_{i-1} + 8U_{i+1} - U_{i+2})/12h. \quad (66c)$$

Calculation of the derivatives using the right differences:

$$E_i = U_i' = E_i^{RC} = (-3U_{i-1} - 10U_i + 18U_{i+1} - 6U_{i+2} + U_{i+3})/12h, \quad (66d)$$

$$E_i = U_i' = E_i^R = (-25U_i + 48U_{i+1} - 36U_{i+2} + 16U_{i+3} - 3U_{i+4})/12h. \quad (66e)$$

It should be noted that the calculations using the central differences are the most accurate.

If a point is located inside the electrode, the potential of this point is assumed to be equal to the potential of the entire electrode, while the field strength at that point equals zero. In the frame of the algorithm proposed, each point belonging to a 2D set of potentials is marked either as the interior or exterior point relative to the system of electrodes under consideration. The algorithm of determining whether a given point belongs to a certain electrode of the entire system consists of the following details.

Consider a system consisting of two electrodes, *Electrode* 1 and *Electrode* 2 (Figure 9), the boundary of which represents a set of straight line segments. The ray OA is directed to the right from a certain *i*th point designated as O.

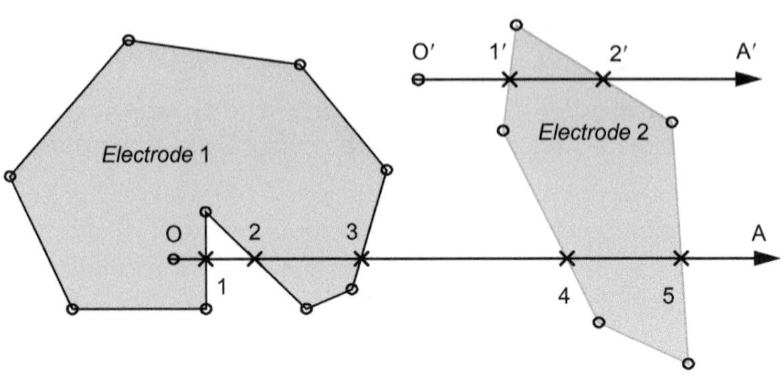

Figure 9 The definition of assigning a point to the electrode.

After that, all points of intersection of the ray OA with the segments belonging to the system's boundary are to be found. In Figure 9, these are the points 1, 2, 3, 4, and 5. If the number of intersection points is odd, the *i*th point is considered to be located inside the electrode, the value «1» ($P_i=1$) is ascribed to that point, and the point's potential is set to be equal to the potential of the electrode with the point of intersection being closest to the point O.

If the number of intersection points is even, as for the ray O′A′ in Figure 9, this means that the *i*th point (point O′) is located outside the electrode, and the value «0» ($P_i=0$) is ascribed to that point.

The essence of the derivative evaluation algorithm based on the use of the shifted finite-difference formulas is that the differences on the right side are employed when approaching the electrode surface from the left, and vice versa. The algorithm is quite simple and effective if we dispose of the information for each of the nodes of the 2D mesh of potentials whether or not a given node belongs to a certain electrode of the system under consideration.

Using Eq. (66) and the data on the location of the points $r_i=ih$, $i=0,1,$ $2 \ldots N$ within any electrode ($P_i=0$ or $P_i=1$), the algorithm for derivative calculation can be arranged in the following ways:

- If the *i*th point is located within the electrode ($P_i=1$), then the field intensity $E_i=0$.
- If the points adjacent to the *i*th point are located outside the electrode (i.e.,$P_{i-2}+P_{i-1}+P_{i+1}+P_{i+2}=0$), the central differences [Eq. (66c)] are used to evaluate the field intensity at the *i*th point: $E_i=E_i^C$.
- If only left adjacent points are located inside the electrode [i.e., $P_{i-2}+P_{i-1}=2$ (Figure 10a)], the right-side differences [Eq. (66e)] are used: $E_i=E_i^R$.
- If only a single terminal point of the two points adjacent to the *i*th point is located inside the electrode [i.e., $P_{i-2}=1$ (Figure 10b)], the right-side differences [Eq. (66d)] are used: $E_i=E_i^{RC}$.
- If only the right-side adjacent points are located inside the electrode [i.e. $P_{i+2}+P_{i+1}=2$ (Figure 10c)], the left-side differences [Eq. (66a)] are used: $E_i=E_i^L$.
- If only a single terminal point of the two right-side points adjacent to the *i*th point is located inside the electrode [i.e., $P_{i+2}=1$ (Figure 10d)], the right-side differences [Eq. (66b)] are used: $E_i=E_i^{LC}$.

Numerical estimates of the relative errors as applied to the abovementioned approximations of derivatives appear as follows: average error is 7.44×10^{-7}; maximum error is 6.39×10^{-5} (see Figure 11).

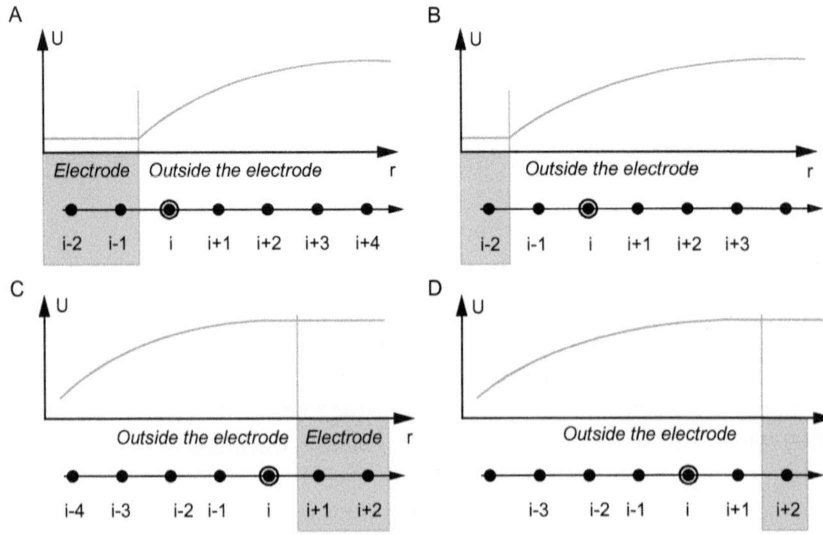

Figure 10 Potential gradient calculation using the right-side (A, B) and left-side (C, D) differences.

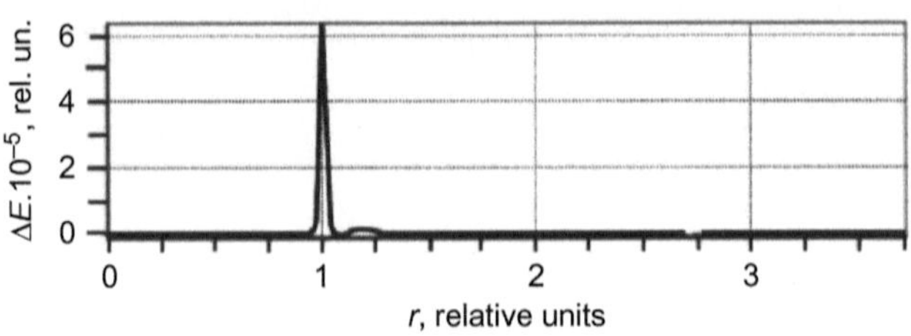

Figure 11 Differentiation accuracy using the shifted finite-difference formulas. (See the color plate.)

6.3 Investigating the Accuracy of Gradient Calculation Using DFT

It may be assumed that a better accuracy of numerical differentiation can be achieved with the use of the discrete Fourier transform (DFT), since the DFT takes advantage of all the information about the function values at all mesh nodes. To calculate derivatives, we use one of the well-known properties of the Fourier transform—namely, the fact that the Fourier transform of the derivative $<dU/dr>$ of a function $U(r)$ can be obtained as a product of the imaginary value iw and the Fourier transform $<U(r)> = \boldsymbol{F}(w)$ of the original function, so that

$$< dU/dr >= iw < U(r) >\rightarrow iw\boldsymbol{F}(w), \qquad (67)$$

where w is the frequency of harmonic signal representation and i is the imaginary unit. The derivative of the function dU/dr is obtained by the inverse Fourier transform applied to the Fourier transform of the derivative $<dU/dr>$.

In the trigonometric form of the DFT, the Fourier transform $F_j = F(w_j)$ of the discrete function $U_k = U(r_k)$, $k = 0,1\ldots N$ is expressed in terms of the amplitudes of sinusoids (the imaginary part of the Fourier transform) $B_j = \frac{1}{m}\sum_{k=0}^{N} U_k \sin\frac{\pi jk}{m}$ and the amplitudes of cosinusoids (the real part of the Fourier transform) $A_j = \frac{1}{m}\sum_{k=0}^{N} U_k \cos\frac{\pi jk}{m}$, where $F_j = A_j + i\,B_j$, $m = N/2$.

Thus, the Fourier transform $\mathbf{G}$ of the derivative $E = dU/dr$ appears as

$$G_j = iw_j\left(A_j + iB_j\right) = -w_j B_j + iw_j A_j = C_j + iD_j, \tag{68}$$

where $C_j = -w_j B_j$ is the real part of $\mathbf{G}$ and $D_j = w_j A_j$ is the imaginary part of $\mathbf{G}$.

Now, according to the formulas of the inverse discrete Fourier transform, we can restore the derivative E_k itself:

$$E_k = \frac{C_0}{2} + \sum_{j=1}^{m-1}\left[C_j\cos\left(k\frac{\pi j}{m}\right) + D_j\sin\left(k\frac{\pi j}{m}\right)\right] + \frac{C_m}{2}\cos(k\pi), \quad k = 0, 1, \ldots, N;\, m = (N+1)/2. \tag{69}$$

The results of applying the DFT to differentiating the test function U are shown in Figure 12. Numerical estimates of relative errors of the approximations considered here are as follows: the average error of derivative calculation is 2.53×10^{-2} and the maximum error is 0.43.

Thus, the result of applying the DFT for calculating the derivatives of the functions having a kink looks not very promising. Our experiments indicate that far more advanced results may be expected from the use of the left-side,

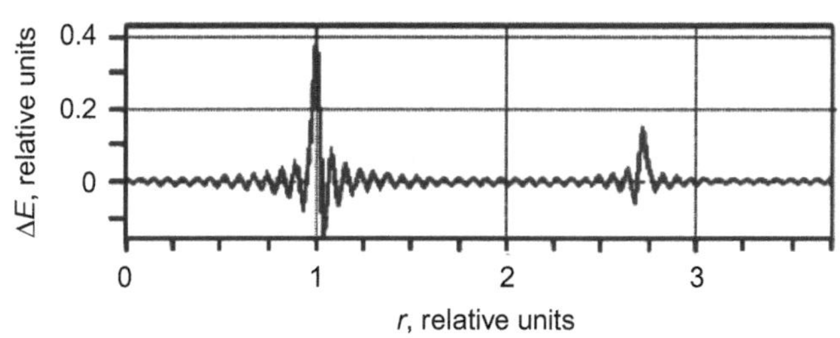

Figure 12 Differentiation accuracy using the DFT. (See the color plate.)

center, and right-side differences, the selection of which is governed by the algorithm highlighted in this discussion.

6.4 2D Interpolation of Potential Gradient

The methods considered in the previous section allow us to determine the x- and y-components of the potential gradient at the nodes of a 2D mesh constructed in the calculation domain (Figure 13). However, numerical integration of the motion equations requires knowledge of the potential gradient components at any arbitrary point located between the nodes. Thus, a 2D interpolation of the nodal values of the gradient components is needed.

In the case of rectangular mesh, the sequential interpolation turns out to be the most convenient. Consider the gradient components $E_x(x, y), E_y(x, y)$ defined on the mesh nodes x_i, y_j. Since the implementation of 2D interpolation is completely identical for both of these components, we introduce a particular mesh function $E_{ij} = E(x_i, y_j)$ to restore the function $E(x,y)$ at the required points of the calculation domain.

Let us select on this mesh a rectangle with $k \times k$ nodes (Figure 13; $k = 4$), containing a point of interest with the coordinates (x,y). This point is designated in Figure 13 by a small white circle. Within the interior part of the calculation domain, the numbers of mesh nodes located on the right and left, as well as above and below, with respect to the desired point are set to be equal. When approaching the boundary, these numbers of mesh nodes becomes naturally different. Note that the node being the closest to the point (x,y) from the left and from below is designated as (i, j).

The practice shows that splines possess excellent interpolation properties for a wide class of functions. In the course of sequential interpolation,

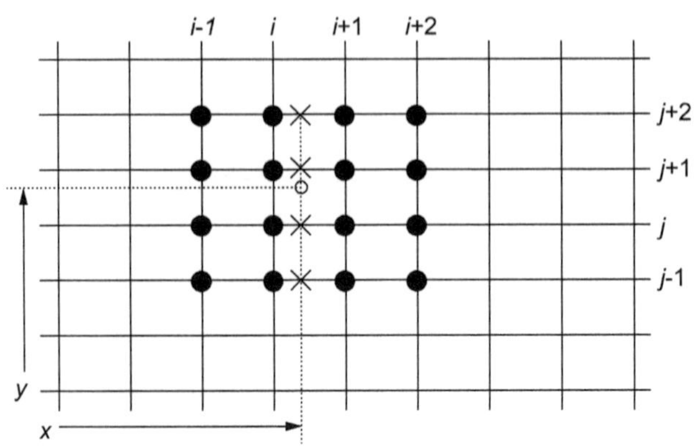

Figure 13 2D interpolation of the potential gradient.

a one-dimensional (1D) spline-interpolation of the rows is being first performed. With this aim in view, the values $E_{j-1}(x)$, $E_j(x)$, $E_{j+1}(x)$, and $E_{j+2}(x)$ of the function E at the points having the coordinate x (designated by crosses) in each of k horizontal rows are determined in accordance with the values E_{i-1m}, E_{im}, E_{i+1m}, E_{i+2m}, and $m = j - 1 \ldots j + 2$. The vertical column of these values allows us to perform spline interpolation of the function E at the point with the coordinate (x, y) and thus determine the desired value of $E(x, y)$.

6.5 3D Interpolation of Axisymmetric Potential Gradient

In simulations of the energy analyzers with off-axis sources, in studies on the influence of external magnetic fields on the parameters of energy analyzers, and in many other problems of charged particle optics, it is necessary to carry out calculations in a Cartesian coordinate system, which requires knowledge of relevant components of potential gradient. In axisymmetric fields, the 2D interpolation algorithm described earlier in this chapter allows reconstruction of the axial and radial gradient components E_x, E_r at any arbitrary point (x, r) of the meridian plane xOr containing the symmetry axis. The transition to Cartesian coordinates is quite easy. At each point (y, z) of the plane yOz perpendicular to the symmetry axis Ox, the y- and z-gradient components can be found as follows (Figure 14):

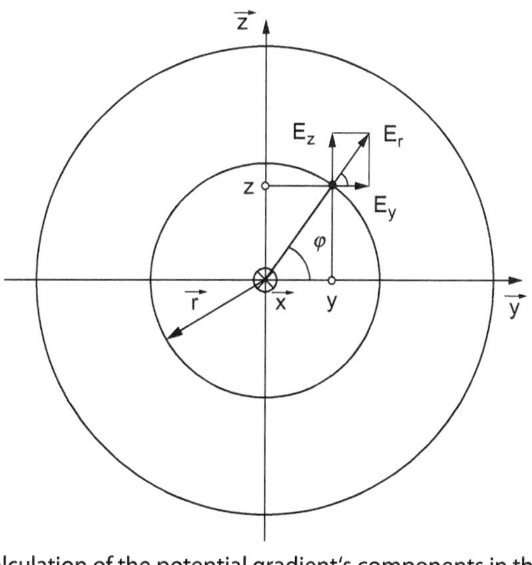

Figure 14 The calculation of the potential gradient's components in the Cartesian coordinate system *Oxyz;* the symmetry axis *Ox* is perpendicular to the plane shown here.

$$E_y(y, z) = E_r(r) \cdot \cos\varphi = E_r(r) \cdot y/r, \tag{70a}$$
$$E_z(y, z) = E_r(r) \cdot \sin\varphi = E_r(r) \cdot z/r, \tag{70b}$$

where $r = \sqrt{y^2 + z^2}$, and the component E_x does not depend on the angle φ and represents a function of the variables in the meridian plane; i.e., $E_x = E_x(x, r)$.

Some specific details of solving the planar Dirichlet problem and calculating the potential gradient in a planar domain are given in Appendix 1 of this book.

7. TRAJECTORY ANALYSIS TECHNIQUES IN CORPUSCULAR-OPTICAL SYSTEMS

The charged particle motion in an electromagnetic field is described by the ordinary second-order differential equations:

$$\begin{aligned} m\frac{d^2x}{dt^2} &= qE_x + q\left(v_y B_z - v_z B_y\right), \\ m\frac{d^2y}{dt^2} &= qE_y + q\left(v_z B_x - v_x B_z\right), \\ m\frac{d^2z}{dt^2} &= qE_z + q\left(v_x B_y - v_y B_x\right), \end{aligned} \tag{71}$$

where v_x, v_y, v_z, E_x, E_y, E_z, B_x, B_y and B_z are the components of the particle's velocity and electric and magnetic fields; q and m are the charge and mass of the particle.

Numerical methods represent the most general and widely spread way of constructing the charge particle trajectories. Currently, one of the methods of numerical integration of the motion equations in charged particle optics that is most common and well acknowledged in practice is the Runge-Kutta method and its various modifications (Forsyte, Malcolm, & Moler, 1977).

An efficient tool of practical calculations is the Runge-Kutta-Fehlberg algorithm, which assumes six calculations of the right side of the differential equation per one integration step, in contrast to four calculations required by the standard fourth-order Runge-Kutta method. Herewith, this method provides a higher (fifth) accuracy order. The main advantage of the method is that the additional calculations of the right side of the equation being integrated turn out to be very useful. By combining four of six calculations, German mathematician Erwin Fehlberg was able to construct one more integral of fourth-order accuracy, the comparison of which with the resulting fifth-order solution ensured a reliable estimation of the integration step.

The accuracy control and setting the integration step are performed in the following way. If the calculation error ε on a particular integration step exceeds the predetermined error ε_0, the integration step is halved, and the integration is repeated from the previous point with a new integration step. The step is decreased until the condition $\varepsilon \leq \varepsilon_0$ is reached. The calculations are continued with the new integration step; however, in the case of a strong decrease of the calculation error ($\varepsilon < \varepsilon_0/100$), the step is increased by half. This procedure realizes the automatic control of the integration step in the domains with different levels of field smoothness, which provides high-quality solutions and, simultaneously results in a reduction in calculation time.

Testing the algorithm in the problems with analytical solutions have demonstrated an increase in the calculation accuracy by more than an order of magnitude and about fivefold to tenfold increase in the average calculation rate compared to the standard Runge-Kutta method, with a guaranteed level of calculation error. This indicates the fact that the Runge-Kutta-Fehlberg method can be effectively used for trajectory analysis in many sophisticated problems of electron and ion optics.

8. CORRELATION METHOD FOR SEEKING THE CONDITIONS OF HIGHER-ORDER ANGULAR FOCUSING

The order of focusing is one of the most important characteristics in electron optics. As applied to electron lenses, this parameter determines the value of spherical aberration, while for dispersive energy-and mass-analyzers, it reflects the degree of a certain contradiction between the achievement of high transmittance and resolution. The numerical method for seeking the conditions of higher-order angular focusing outlined in this section (see also Trubitsyn, 2000) can be considered as a development of the method presented in Gorelik (1986) and Gorelik, Protopopov, and Trubitsyn (1988). The essence of the method consists of the following details.

After a charged particle has passed through a dispersive or focusing field, the ordinate $y(\mathbf{k}, \alpha)$ of the point of intersection of its planar trajectory with a straight line perpendicular to the Ox-axis and located at a distance x from the source of charged particles can be expanded into the Taylor series:

$$y(\mathbf{k}, \alpha) = y(\mathbf{k}, \alpha_0) + (\partial y/\partial \alpha)_0 \Delta\alpha + 1/2\left(\partial^2 y/\partial \alpha^2\right)_0 \Delta\alpha^2 + \ldots. \quad (72)$$

where α is the angle of particle's emission from the source, $\Delta\alpha = (\alpha\text{-}\alpha_0)$, and **k** is the vector, whose components represent the distance x, the particle energy E, and the dimensions and electrode potentials of an electron-optical system. In further calculations, **k** is omitted. In accordance with the Taylor series, the condition of Nth-order focusing for a given **k** can be expressed as vanishing relevant partial derivatives with respect to the angle α:

$$y'(\alpha_0) = y''(\alpha_0) = \ldots = y^{(N)}(\alpha_0) = 0, \tag{73}$$

In accordance with the method proposed in Gorelik et al. (1988), the function $y(\alpha)$ can be represented as

$$y(\alpha) = R(\alpha)\text{-}x \times t(\alpha), \tag{74}$$

where $R(\alpha) = y_c(\alpha) + x_c(\alpha) \times t(\alpha)$, $t(\alpha) = \text{tg}(\gamma)$, $\gamma(\alpha)$ and $x_c(\alpha)$, $y_c(\alpha)$ are, respectively, the angle relative to the axis $0x$ and coordinates of the particle escape from the area with a nonzero field gradient, which corresponds to the time moment of crossing the output electrode plane.

In accordance with Eq. (74), the condition of first-order focusing $y'(\alpha) = 0$ takes the form

$$R'(\alpha)\text{-}x\,t'(\alpha) = 0 \tag{75}$$

or

$$x = R'(\alpha)/t'(\alpha),\ y = R(\alpha) - x \times t(\alpha). \tag{76}$$

The second derivative $y''(\alpha)$ vanishes if

$$R''(\alpha)\text{-}x \cdot t''(\alpha) = 0, \tag{77}$$

and, with regard to the expression for x in Eq. (76), the condition of second-order focusing can be written as

$$R''(\alpha)\,t'(\alpha)\text{-}R'(\alpha)\,t''(\alpha) = 0. \tag{78}$$

After that, the function

$$F(\alpha) = R''(\alpha)\,t'(\alpha) - R'(\alpha)\,t''(\alpha) \tag{79}$$

is introduced and a search for the second-order focusing conditions is performed by solving the equation

$$F(\alpha) = 0 \tag{80}$$

relative to $\alpha = \alpha_0$ and determining the points of focusing

$$x_0 = R'(\alpha_0)/t'(\alpha_0),\ y_0 = R(\alpha_0)\text{-}x_0 t(\alpha_0). \tag{81}$$

When seeking the focusing conditions numerically, the required functions are calculated on a discrete set of input angles α_i, $i = 1, 2, \ldots, L$ (the amount of 20÷50 trajectories is commonly sufficient) located in a previously selected range of $[\alpha_{min}, \alpha_{max}]$ and interpolated for $\alpha \neq \alpha_i$. The derivatives are calculated according to the formulas of numerical differentiation.

It follows from Eqs. (74) and (76) that the condition of Nth-order focusing is fulfilled if

$$\delta_n(\alpha_0) = R^{(n)}(\alpha_0)\cdot t'(\alpha_0)\text{-}R'(\alpha_0)\cdot t^{(n)}(\alpha_0) = 0, \tag{82}$$

where $n = 2, 3, \ldots, N$. However, numerical calculation of higher derivatives $(n > 2)$ leads to large errors, so the function $\delta_n(\alpha)$ cannot be calculated with the desired accuracy.

Experience shows this method to be effective in seeking the conditions of second-order focusing, but hardly suitable for constructing the focusing conditions of higher orders. This raises the problem of extending this method to the search of focusing conditions having an order higher than 2.

Eq. (82) says that in the case of Nth-order focusing for $2 \leq m \leq N$,

$$R'(\alpha_0)/t'(\alpha_0) = R^{(m)}(\alpha_0)/t^{(m)}(\alpha_0), \tag{83}$$

and for $m = n - 1$, Eq. (82) may be rewritten as

$$\delta_n(\alpha_0) = R^{(n)}(\alpha_0)\cdot t^{(n\text{-}1)}(\alpha_0)\text{-}R^{(n\text{-}1)}(\alpha_0)\cdot t^{(n)}(\alpha_0) = 0, \tag{84}$$

where $n = 2, 3, \ldots, N$.

Eq. (79) determines the derivative $F'(\alpha)$:

$$F'(\alpha) = R'''(\alpha)\cdot t'(\alpha) + R''(\alpha)\cdot t''(\alpha)\text{-}t''(\alpha)\cdot R''(\alpha)\text{-}t'''(\alpha)\cdot R'(\alpha) =$$
$$= R'''(\alpha)\cdot t'(\alpha)\text{-}t'''(\alpha)\cdot R'(\alpha). \tag{85}$$

It follows from Eq. (82) that $F'(\alpha_0) = 0$ in the case of third-order focusing. The penultimate equation gives the equality

$$F''(\alpha) = R^{(IV)}(\alpha)\cdot t'(\alpha) + R'''(\alpha)\cdot t''(\alpha)\text{-}t^{(IV)}(\alpha)\cdot R'(\alpha)\text{-}t'''(\alpha)\cdot R''(\alpha). \tag{86}$$

According to Eq. (84), in the case of third-order focusing,

$$F''(\alpha_0) = R^{(IV)}(\alpha_0)\cdot t'(\alpha_0) - t^{(IV)}(\alpha_0)\cdot R'(\alpha_0), \tag{87}$$

while for fourth-order focusing, $F''(\alpha_0) = 0$ [see Eq. (82)].

Similar considerations show that the focusing condition of the $(N+2)$th order appears as

$$F^{(n)}(\alpha_0) = 0,\ n = 0, 1, 2, \ldots, N. \tag{88}$$

Let us expand the function $F(\alpha)$ in the vicinity of α_0:

$$\begin{aligned} F(\alpha) = F(\alpha_0) + F'(\alpha_0)(\alpha\text{-}\alpha_0) + (1/2)F''(\alpha_0)(\alpha - \alpha_0)^2 + \ldots \\ + (1/N!)F^{(N)}(\alpha_0)(\alpha\text{-}\alpha_0)^N + \ldots \end{aligned} \tag{89}$$

Taking into consideration Eq. (88), in the case of focusing order $(N + 2)$, we obtain

$$\begin{aligned} F(\alpha) = 1/(N+1)!\, F^{(N+1)}(\alpha_0)\,(\alpha\text{-}\alpha_0)^{N+1} \\ + 1/(N+2)!\, F^{(N+2)}(\alpha_0)\,(\alpha\text{-}\alpha_0)^{N+2} + \ldots \end{aligned} \tag{90}$$

Since the function $F(\alpha)$ is calculated from Eq. (79) numerically, it represents a superposition of true values and certain noise. Therefore, this function should be treated as a random function dependent on a nonrandom argument α.

Denote $S^*(\alpha)$ as the function determined by Eq. (90) as follows:

$$\begin{aligned} S^*(\alpha) &= 1/(N+1)!\, F^{(N+1)}(\alpha_0)\,(\alpha\text{-}\alpha_0)^{N+1} \\ &\quad + 1/(N+2)!\, F^{(N+2)}(\alpha_0)\,(\alpha\text{-}\alpha_0)^{N+2} + \ldots \\ &\approx 1/(N+1)!\, F^{(N+1)}(\alpha_0)\,(\alpha\text{-}\alpha_0)^{N+1}. \end{aligned} \tag{91}$$

We also consider $S^*(\alpha_0)$ as a random function with a negligibly low noise level and the nonrandom multiplier $1/(N+1)!\, F^{(N+1)}(\alpha_0)$. It is rather obvious that in the case of $(N + 2)$th-order of focusing, the functions $F(\alpha)$ and $S^*(\alpha)$ are correlated.

To evaluate the degree of dependence between the cross sections of two random functions, the normalized cross-correlation function is commonly used (Bendat & Piersol, 1971). Since multiplication of a random function by the nonrandom factors does not change the normalized cross-correlation function, it is enough to explore the correlation between $F(\alpha)$ and $S(\alpha) = (\alpha\text{-}\alpha_0)^{N+1}$ in order to evaluate the correlation dependence between the functions $F(\alpha)$ и $S^*(\alpha)$.

In the case of discrete change of the argument $\alpha = \alpha_1, \alpha_2, \ldots, \alpha_L$ and zero shift between the functions $F_i = F(\alpha_i)$ and $S_i = S(\alpha_i) = (\alpha_i - \alpha_0)^{N+1}$, $i = 1,2, \ldots, L$, the normalized cross-correlation function is defined as

$$\rho_0(N) = R_{FS}/(K_F \cdot K_S)^{1/2}, \tag{92}$$

where $R_{FS} = \frac{1}{L}\sum_{i=1}^{L}(F_i - \bar{F})(S_i - \bar{S})$, $K_F = \frac{1}{L}\sum_{i=1}^{L}(F_i - \bar{F})^2$, $K_S = \frac{1}{L}\sum_{i=1}^{L}(S_i - \bar{S})^2$, $\bar{F} = \frac{1}{L}\sum_{i=1}^{L} F_i$, $\bar{S} = \frac{1}{L}\sum_{i=1}^{L} S_i$. The higher the degree of correlation between the functions F_i and S_i is, the closer the function ρ_0 is to the unit. Here, we consider ρ_0 as a function of the parameter N that determines the power of the relevant polynomial.

The approach that we suggest for seeking the conditions of higher-order focusing assumes the following. The angle α_0 of second-order focusing is determined from Eq. (80). After that, the correlation between the function $F(\alpha)$ and the power function $S(\alpha) = (\alpha\text{-}\alpha_0)^{m+1}$, with parameter m taking sequentially the values $m = 0, 1, \ldots, M$, is estimated using Eq. (92). Here, M is the upper limit of the search, which is selected in accordance with a particular problem under consideration.

Then we determine the value N $(0 \leq N \leq M)$ for which $\rho_0(N) = \max \{\rho_0(0), \rho_0(1), \ldots, \rho_0(M)\}$. The proximity $\rho_0(N)$ to 1 indicates the cross-correlation of $F(\alpha)$ and $S(\alpha) = (\alpha\text{-}\alpha_0)^{N+1}$; i.e., the presence of the focusing of the $(N + 2)$th order. In practice, it suffices to consider $M = 10 \div 20$. To avoid operations with large numbers, the function $S(\alpha) = (\alpha\text{-}\alpha_0)^{N+1}$ should be normalized by the factor $(\alpha_{max}\text{-}\alpha_{min})^{N+1}$.

It follows from Eq. (90) that, at high orders of focusing, the function $F(\alpha)$ is close to zero in a fairly wide range of angles α. Therefore, a noise in $F(\alpha)$ may result in the appearance of the false roots when solving the equation $F(\alpha) = 0$. In such cases, in order to calculate the α_0 value more precisely, it is necessary to evaluate the maximum of the cross-correlation function $\rho_0(\alpha_0, N)$ in two variables—the central angle of focusing and the polynomial degree. Next, we present the results of testing this method by means of a model admitting an analytical solution.

As has been established (Zashkvara, Korsunskii, & Kosmachev, 1966), cylindrical mirrors possess the property of second-order angular focusing near $\alpha_0 \approx 42°$. The trajectory analysis of cylindrical mirror conducted in accordance with the procedure described by Trubitsyn (1995b) indicates that the second-order focusing at $\alpha_0 \approx 42°$ may also be implemented if the fringe field in a real device is corrected via three pairs of adjusting rings. Herewith, the corresponding value of the normalized cross-correlation function for $F(\alpha)$ and $(\alpha\text{-}\alpha_0)^{m+1}$ attains its maximum at $m = 0$ and equals $\rho_0 = 0.99$.

The energy analyzer with three cylindrical electrodes (Franzen & Taaffe, 1983) provides third-order focusing at the central angle $\alpha_0 \approx 40°$. The correlation function ρ_0 reaches its maximum value of 0.97 for $m = 1$ in the case of numerical integration of the charged particle trajectories in an analytically defined electrostatic field. The focus point's coordinates coincide with the coordinates that have been determined analytically by Franzen and Taaffe.

Trajectory analysis of the spherical mirror ensuring ideal angular focusing (Zashkvara, Yurchak, & Bylinkin, 1988) at $\alpha = 90°$, as well as the estimate $\rho_0 > 0.95$ that we obtained for the correlation function at the polynomial degree $N + 1 = 25$ (this is maximal polynomial degree used in our calculations), confirm the efficiency of the abovementioned technique in exploring the electron-optical systems with a high focusing level.

9. EXAMPLES OF NUMERICAL SIMULATION OF ENERGY ANALYZERS

This section discusses the results of numerical simulation of energy analyzers with different configuration of electrodes on the basis of the numerical techniques considered previously.

9.1 Systems Based on Cylindrical Optics

The instruments based on cylindrical optics, thanks to the ease of manufacture and capability of providing a high level of luminosity, are most commonly used for transportation, focusing, and energy analysis of the charged particle flows.

9.1.1 High-Resolution Energy Analyzer

The cylindrical mirror analyzer (CMA) (Zashkvara et al., 1966) is one of the most used electron spectrometers, which is conditioned by a number of its advantages, such as high energy resolution at a considerable transmittance, design simplicity, and the opportunity of employing a coaxially embedded electron gun. The limited size of real instruments requires reducing the fringe effects. In general, complete elimination of edge perturbations in an electrostatic field is impossible, and this is the reason why the characteristics of actual designs may differ from those of idealized models.

The order of angular focusing represents one of the main criteria of the energy analyzer quality. The higher the order of angular focusing is, the weaker the contradiction is between the luminosity and resolution. This raises the problem of studying the focusing properties of instruments with taking into consideration the fringe effects. One way to reduce the fringe effects assumes the use of several pairs of adjusting rings being insulated from each other and having the following potentials proportional to the potential V_b of the external cylinder (the potential V_a of the inner cylinder is assumed to be equal to zero): $u_i = iV_b/(n+1)$, where i is the sequential number of a ring, counted from the inner cylinder, and n is the number of pairs of adjusting rings. The internal (r_i^{in}) and external (r_i^{out}) radii of the ith correcting ring are calculated according to the formulas

$$r_i^{in}/r_a = 0.5\left[(r_b/r_a)^{u_i} + (r_b/r_a)^{u_{i-1}} + d/r_a\right],\ r_i^{out} = r_{i+1}^{\ in} - d, \qquad (93)$$

where $i = 1,2 \ldots n$, $u_0 = V_a$ and $u_{n+1} = V_b$, r_a and r_b are the radii of internal and external cylinders, and $d \geq 0$ is the insulating gap between the correcting electrodes.

A schematic of such a device, with two pairs of adjusting rings, is shown in Figure 15.

In actual design, the following recommendation (Zashkvara, Ashimbaeva, Chokin, & Rysavy, 1992) is to be taken into account: the apertures of the input and output windows must be closed by 1D grids consisting of the segments of wire spanned along the inner cylinder surface and

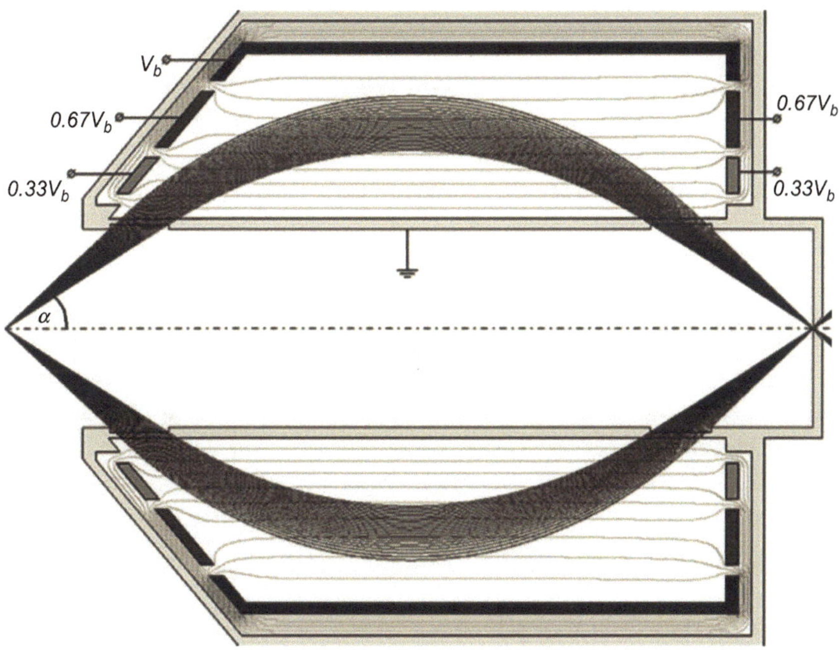

Figure 15 Transaxial cross section of the energy analyzer, with two pairs of correcting rings, and electron trajectories with the energy E_0 ($E_0/V_b = 1.4311$). (See the color plate.)

separated by a certain angular interval. The advantage of the 1D grid is that such a grid does not affect the focusing in the axial plane and scatters the electron beam in the azimuthal direction only.

Numerical calculations have shown that the CMA with three, four, and five pairs of correcting rings provide the mode of second-order angular focusing that corresponds to the ideal cylindrical mirror. The focusing angle, focus position, and energy dispersion in the CMA with three pairs of correcting rings differ by a few percentage points from the corresponding parameters of the idealized device. This difference decreases with increasing the number of correcting electrodes.

Of greatest interest are the results obtained for the energy analyzer with two pairs of correcting rings; i.e., for the case that the energy analyzer's field considerably differs from the field in the idealized mirror. The corresponding calculations indicate the possibility of fourth-order focusing with a central angle of 42.3°.

The increase in focusing order improves the consumer characteristics of the energy analyzer. Energy resolution and luminosity are the consumer characteristics of energy analyzers in electron spectroscopy. Both of these parameters can be characterized in terms of the instrumental function. Next, we compare the instrumental function of the CMA with second-order angular focusing with the instrumental function of the energy analyzer being proposed here, which offers fourth-order angular focusing.

Figure 16 (curve 1) shows the instrumental function of the energy analyzer with three pairs of correcting rings, which provides second-order focusing, as noted previously. The output aperture, with a width of $0.017r_a$, is placed at the focal point. The optimal range of angles is 37°–48°. The analyzer luminosity is 6.5% and the basic energy resolution $R_B = 1.14\%$, while the relative full-width half-maximum (FWHM) of the instrumental function amounts to 0.55%.

Figure 16 (curve 2) represents the instrumental function of the energy analyzer with fourth-order focusing. The output aperture is placed in the plane of minimal cross section of the output electron flow, and its width equals $0.004r_a$.

Comparison of curves 1 and 2 in Figure 16 suggests that a perturbation of the ideal cylindrical field by two pairs of correcting rings provides tenfold improvement in the relative energy resolution (FWHM = 0.05%), while maintaining the luminosity value.

The stability of the instrument parameters to various perturbations is an important practical characteristic. Changing the number of correcting rings placed at the rear edge of the analyzer's cylinders does not affect the focusing

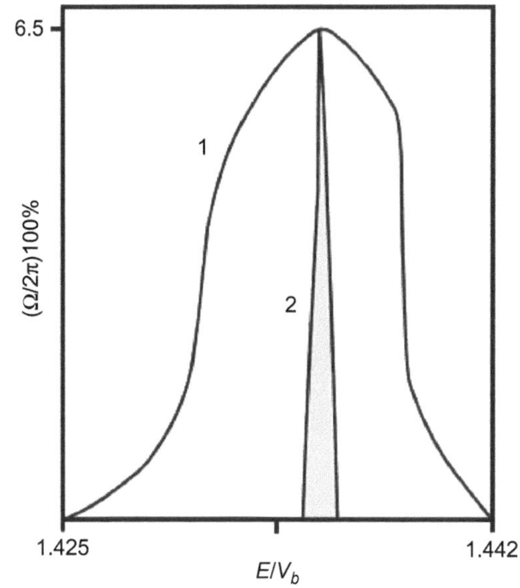

Figure 16 Instrumental function of the energy analyzers with equal luminosity values: 1—second-order focusing mode, 2—fourth-order focusing mode.

order. This means that the main field inhomogeneity is created by the conical correcting electrodes placed at the cylinder's front edge.

The fourth-order focusing mode is conserved when the tilt angle of the section of conical correcting electrodes increases from 50° to 70°. The energy analyzer transition into the mode of second-order focusing when the focusing angle exceeds 70° is explained by the fact that electrostatic field inhomogeneity is eliminated from the area of electron motion. The fourth-order focusing mode is also conserved when changing the focal length (the distance between the source and the front edge of the inner cylinder), which equals $0.8r_a$ and varies within 10% along the z-axis. The analyzer switches to the standard mode when the focal length increase exceeds a specified limit. This fact also is associated with escape of charged particles from the area of field inhomogeneity. Simulations have shown that the condition of fourth-order focusing depends neither on the magnitude of the insulating gap between the electrodes in the range $d = 0\text{-}0.1r_a$ nor on the radii of the inner (r_a) and outer (r_b) cylinders.

9.1.2 Energy Analyzer with Angular Resolution

Simultaneous registration of energy and angular characteristics of secondary electron flows excited by synchrotron radiation allows for the performance of fundamental research in the physics and chemistry of the surface of solids, associated with the problem of adsorption and related phenomena. This includes studying the electronic structure of atoms and molecules adsorbed on the surface and exploring the changes in the energy structure of electronic states of the near-surface layers in solids, as well as acquiring information on localization of atoms on the surface. The progress in this field of science largely depends on the engineering capabilities of relevant scientific instruments, particularly on the parameters of energy analyzers.

Kover et al. (1983) proposed an efficient scheme of a spectrometer with energy/angular resolution, the essence of which is shown in Figure 17. A point source located on the z-axis of a dispersion analyzer emits a fan-shaped flow of charged particles, the median plane of which is perpendicular to the z-axis ($\alpha = \pi/2$). After passing through an electrostatic field, the flow forms an image that is displayed on the position-sensitive detector (PSD). Due to the dispersion properties of the analyzer, the coordinates of the image points depend on the flow energy and the polar angle φ of each electron's emission from the sample. The output aperture of the analyzer "separates" from the output flow the electrons belonging to a certain band of initial energies. The instant analysis of the output flow relative to the polar angle, which is carried out with the use of PSD, and sequential rotation of the probing sample around the normal, allow for exploring the distribution of photoelectrons, whose energies fit the energy settings in the polar angle range of $-\pi/2$ to $\pi/2$ and in the azimuthal angle range of zero to 2π.

This scheme has been employed by a number of researchers in designing analyzers with various configurations of electrodes (see, for example, Varga, Tokesi, & Rajta, 1995 and Leckey, 1987). However, the complexity of the proposed structures raises some doubt as to the possibility of their wide application in practice, so that the problem of development and design of an

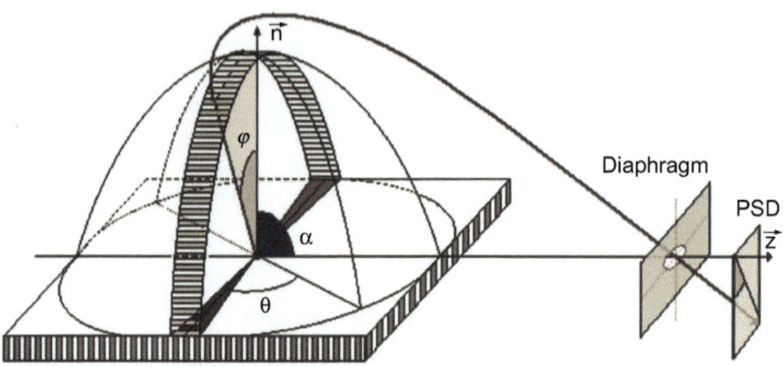

Figure 17 Schematic of angular measurements. (See the color plate.)

analyzer with a simple configuration of electrodes, capable of providing the energy dispersion and angular focusing of the flows of secondary particles with an input angle α close to 90°, still remains very urgent.

The most simple and easy-to-manufacture axisymmetric geometric shapes are cylinder and ring. Figure 18 represents an electron-optical scheme of the box-type device being a combination of these shapes. This arrangement ensures fourth-order angular focusing with nonzero energy dispersion.

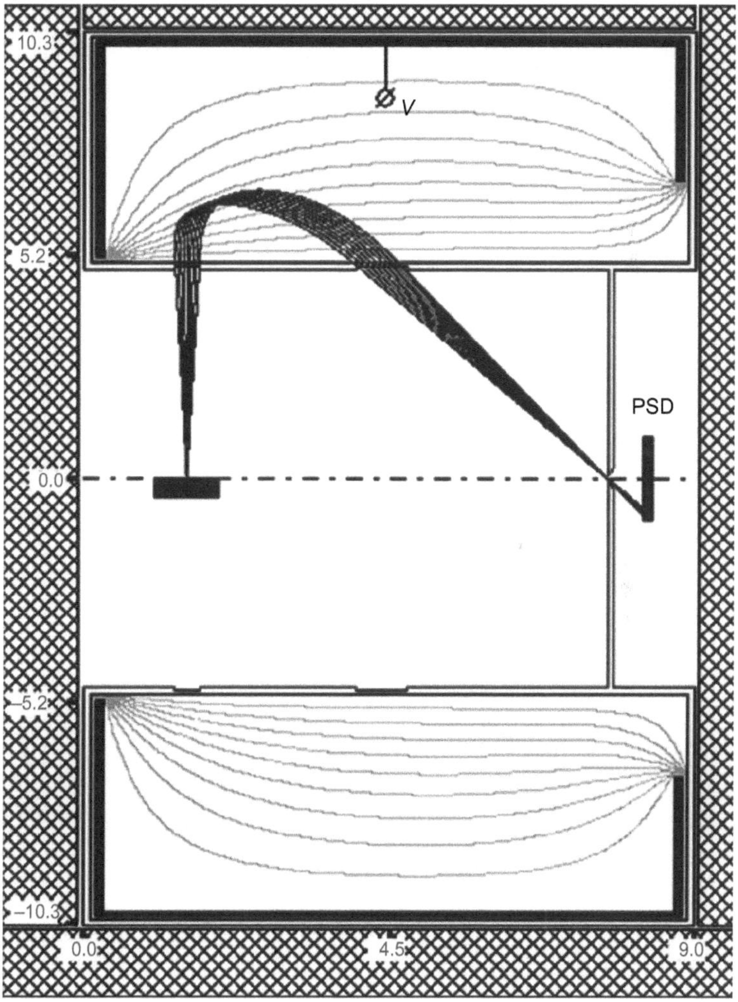

Figure 18 Electron-optical scheme of the analyzer with energy and angular resolution: 1—input aperture, 2—output aperture, 3—position-sensitive detector with selectivity in the polar angle. (See the color plate.)

The fourth-order focusing serves as a criterion of high quality of the energy analyzer, since in this case the contradiction in simultaneous reaching the high levels of luminosity and resolution is considerably weakened compared to the case of lower-order focusing. In the design under consideration, the ratio E/V is 0.655, the ratio of radii of the exterior and interior cylinders is $r_b/r_a = 10/5 = 2$, the distance between the left edge of the analyzer and the source is $0.152r_b$.

The central angle α_0 of the second-order focusing and geometric position of the image depend on the initial electron energy E and the source position on the z-axis. The analyzer's instrumental function is shown in Figure 19. Energy resolution at the instrumental function's half-height constitutes 0.1% for the range of input angles α from 88° to 92° and the output aperture radius $0.002r_a$. The luminosity amounts to 3.5%, given that only the upper half of the instrument is operating.

Another remarkable and unique feature of the proposed scheme is its planar focusing region representing a set of second-order annular focuses that correspond to electrons with different initial energies (see Figure 20, displaying only the upper part of the axial cross section).

This feature allows the analyzer to operate in the spectrographic mode with integral angular collection in the range of angles φ from $-\pi/2$ to $\pi/2$; herewith the PSD with selectivity along the annular image's radius can be used as a multichannel registration system. Thus, this device allows implementation of two independent modes of registering the photoelectron

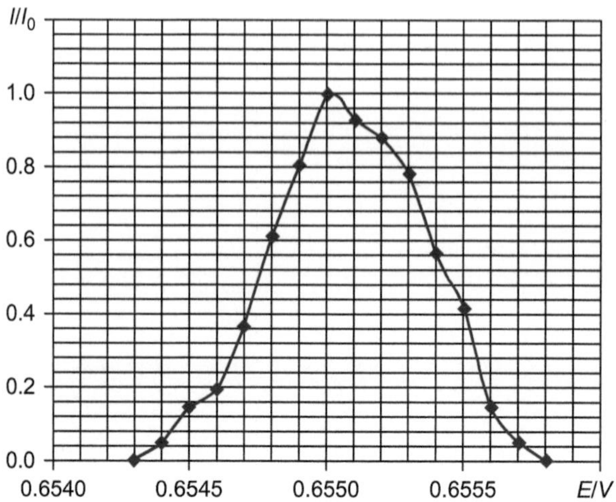

Figure 19 Instrumental function of the analyzer with angular resolution. (See the color plate.)

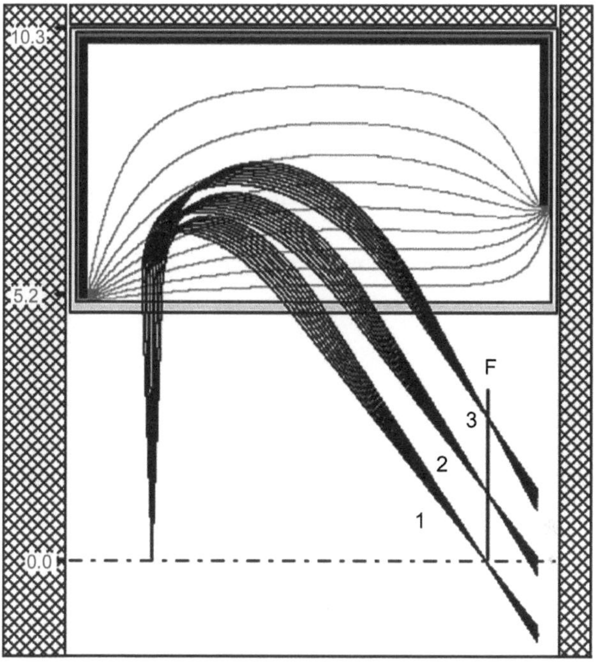

Figure 20 Angular focusing of photoelectrons with different initial energies: 1—$E/V = 0.655$, 2—$E/V = 0.75$, 3—$E/V = 0.85$; F—focal plane. (See the color plate.)

spectra: (i) with angular resolution and (ii) with integral collection of photoelectrons having the emission angles φ in the range from $-\pi/2$ to $\pi/2$ and α in the range of $90° \pm \Delta\alpha$, $\Delta\alpha \geq 2°$. Both modes may be implemented in a single device with independent operation of its lower and upper halves.

Numerical calculations show that small deviations in the geometry of electrodes and initial conditions do not cause any significant changes in the device characteristics.

9.2 Systems Based on Conical Optics

Systems with conical electrodes have a number of useful features and, together with energy analyzers based on cylindrical optics, are widely used in electron spectrometers. This section presents the results of a numerical study on conical electrostatic energy analyzers with spaced and superposed vertices.

9.2.1 Systems with Superposed Vertices

An approximative theory of coaxial conical systems with superposed vertices has been developed by Baranova, Dyakova, and Yavor (1988a). The approach of this work considers energy analyzers in the mode of first-order focusing. As we have already noted, an acceptable level of the energy analyzer characteristics can be reached in the mode of second-order focusing. For this purpose, calculations of the energy analyzer with second-order focusing, point source located on the symmetry axis, and analytical potential distribution defined in spherical coordinates (ρ, θ, ψ) as

$$\varphi(\theta) = V\frac{\ln(\mathrm{tg}(\theta/2)/\mathrm{tg}(\theta_0/2))}{\ln(\mathrm{tg}(\theta_1/2)/\mathrm{tg}(\theta_0/2))}, \tag{94}$$

have been conducted (Figure 21). Here, θ_0 is the half-angle of the inner cone with zero potential, θ_1 is the half-angle of the outer cone having the potential V. Figure 22 shows the angle $\beta_0 = \alpha_0\text{-}\theta_0$, in the vicinity of which the condition of second-order focusing is fulfilled, as a function of the normalized energy of electrons:

$$q = \frac{E}{V}\ln\frac{\mathrm{tg}\theta_1/2}{\mathrm{tg}\theta_0/2}, \tag{95}$$

where β_0 is the angle between the inner cone's generatrix and the particle's trajectory, α_0 is the emission angle of particles, and E is the particle's energy. The dependence of β_0 on q is linear and possesses a relatively high slope that provides more opportunities in the design of this type energy analyzers compared to a cylindrical mirror. For each fixed q, the second-order focusing angle β_0 represents a function of the inner cone's half-angle. This dependence is weak in the range of angles $\theta_0 = 10° \div 40°$. The dashed curve in Figure 12 separates the domains of imaginary (to the left of the curve) and real (to the right of the curve) images. It is seen that the normalized energy q must be increased to obtain a real image with increasing θ_0.

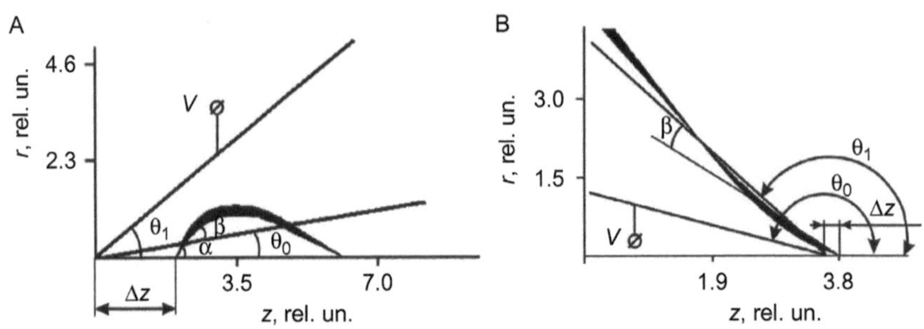

Figure 21 Conical systems with superimposed vertices (the upper half of the system's meridional cross section is shown).

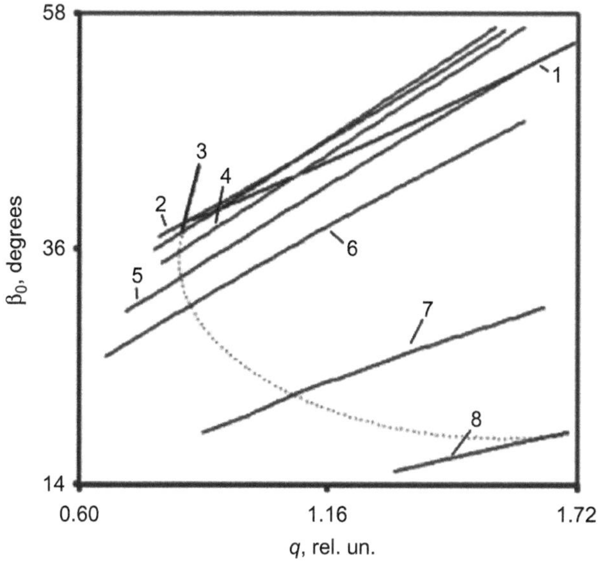

Figure 22 Dependence of the second-order focusing angle on the normalized energy Angles θ_0: 1—10°, 2—20°, 3—30°, 4—40°, 5—50°, 6—60°, 7—90°, 8—110°. (See the color plate.)

The relative linear dispersion D/r increases as q increases and can exceed (by several times) the relevant parameter of the cylindrical mirror $D = E\,\partial r/\partial E$, where r is the distance between the source and the image.

Conical energy analyzers with a moderate value of θ_0 (up to 30°) (Figure 21a) possess a small image radius and high relative linear dispersion (in particular, $D/r \cong 5$ for $q = 1.4$ and $\theta_0 = 10°$), which makes them promising in terms of multipurpose optimization aimed at providing the required geometry of experiment, high operating characteristics, and design simplicity. Of practical interest is the case with $q = 1.76$ and $\theta_0 = 27°$, for which the first-order focusing angle is $\alpha_0 = 90°$ ($\beta_0 = 63°$) and the output electrons have small radial velocity components.

In the range of input angles $\Delta\alpha = \pm 6°$, the output angular spread is $\pm 1°$, relative coordinates of the focus point satisfy the relations $|z_F - z_S|/\Delta z \cong 54.3$, $r_F/\Delta z \cong 7.2$, where z_F, r_F, z_S are, respectively, the focus and source coordinates in the coordinate system associated with the cone's vertices; and Δz is the distance between the axial source and vertices. These features allow one to employ the energy analyzer for measuring the flow intensity of secondary electrons relative to the polar angles in the range from $-\pi/2$ to $\pi/2$ and azimuthal angles in the range from zero to 2π (see the scheme in Figure 17). Owing to the small spread of the output angles, it is possible to implement a retarding mode in the two-stage version of the device by means of a system of two planar grids located between the first and second stages.

The conical systems with $\theta_0 > 90°$ (Figure 21b), or, in terms of the work (Baranova et al., 1988a), the systems with input of electrons through the upper electrode, are less useful from a practical viewpoint and can be applied as the adjustment stages in multistage designs.

9.2.2 Systems with Parallel Generatrices

The potential distribution in conical systems with separated vertices has no analytic expressions and should be calculated numerically. The trajectory analysis of the systems under consideration (Figure 23), having been performed for the case of the point source location on the symmetry axis, indicates the opportunity of implementing the mode of second-order focusing in a broad range of boundary and initial conditions.

In Baranova, Dyakova, and Yavor (1988b), an approximate analysis of the focusing properties of conical systems with parallel generatrices has been conducted. The conditions of first-order focusing in the case of the considerable distance of the particle source to the cone's vertices are investigated in detail. In particular, it is shown that, at $E/V = \text{const}$, the angle β_0 of first-order focusing decreases as the half-angle θ_0 increases.

Figure 24 represents the dependencies of the second-order focusing angle β_0 on the relative distance $\Delta z/g$ between the source and the inner cone's vertex, where g is the distance between the cone's vertices. Reducing the distance $\Delta z/g$ causes a rather sharp increase in angle β_0, which also occurs

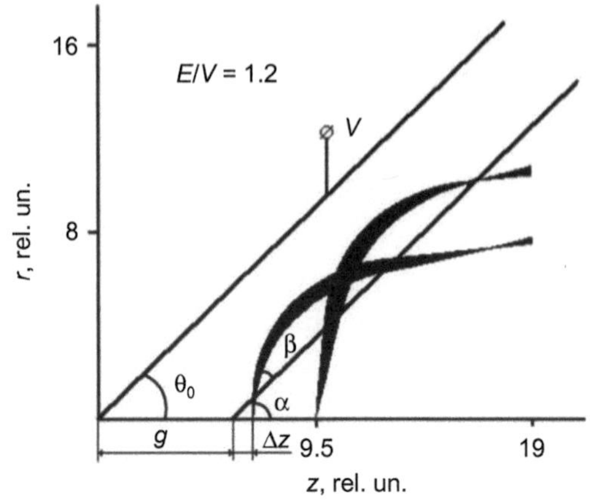

Figure 23 A conical system with parallel generatrices (the upper half of the transaxial cross sections is shown).

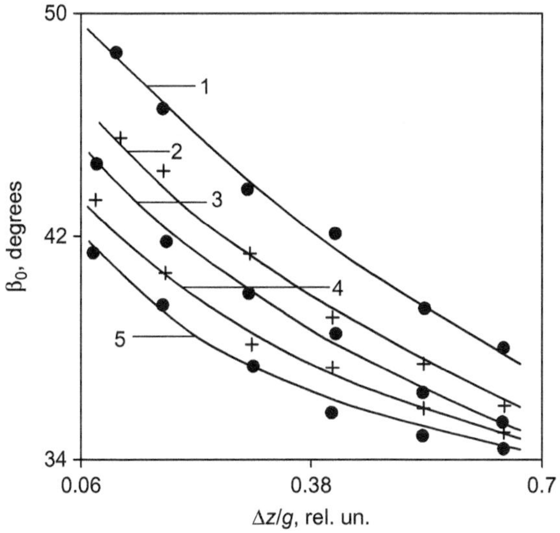

Figure 24 Dependence of the second-order focusing angle on the relative distance between the source and the inner cone's vertex for the following angles θ_0: 1—10°, 2—20°, 3—30°, 4—40°, 5—44°.

when reducing the angular opening of the cones or increasing the relative energy E/V. These features may be useful in designing analyzers capable of providing the desired geometry in the experiment. When the source is moved away at the distance Δz (being much greater than g), the second-order focusing angle β_0 tends toward 30°, which corresponds to the data given by Baranova et al. (1988b).

The calculation shows that relative linear dispersion depends very little on variations of defining parameters and lies within the range of 1–1.5.

The calculation results obtained for the idealized systems (extended cones) reveal the relationship between the parameters providing the second-order focusing and may serve as a criterion for choosing the initial approximations for energy analyzer designs. As an example, Figure 25 represents a scheme of a three-stage analyzer design with the central angle $\alpha_0 =$ 60° that provides the mode of second-order focusing at $E/V = 1$. The first (labeled 1 in the figure) and last (3) device's stages are conical, while the intermediate lens system (2) ensures focusing and transportation of electron beams between stages 1 and 3. According to the results of numerical calculations, the relative energy resolution of the device at the FWHM level of instrumental function (Figure 26) is 0.8% at the luminosity of 3%.

The energy analyzer of this type can be used in the creation of a specialized module for electron spectroscopy of the surface, with a built-in large source of probing (particularly X-ray) radiation.

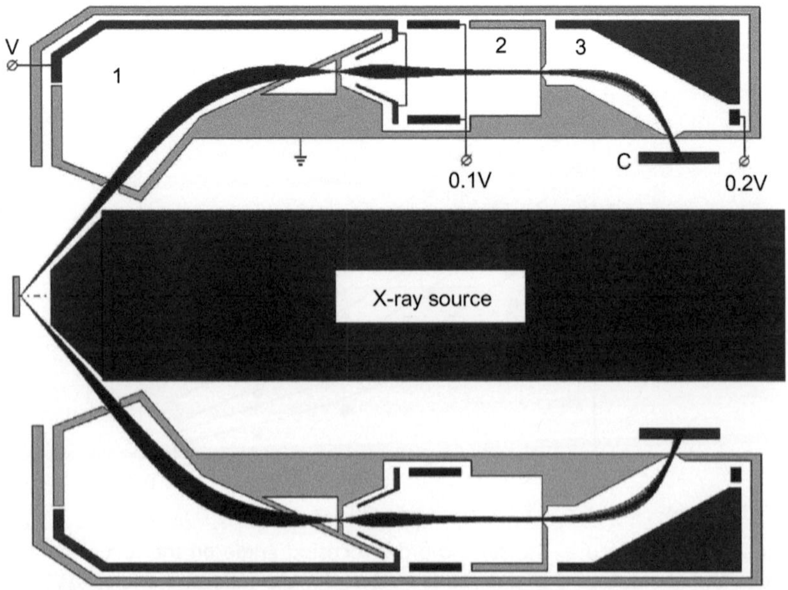

Figure 25 Electron-optical scheme of a three-stage conical energy analyzer with an X-ray source. (See the color plate.)

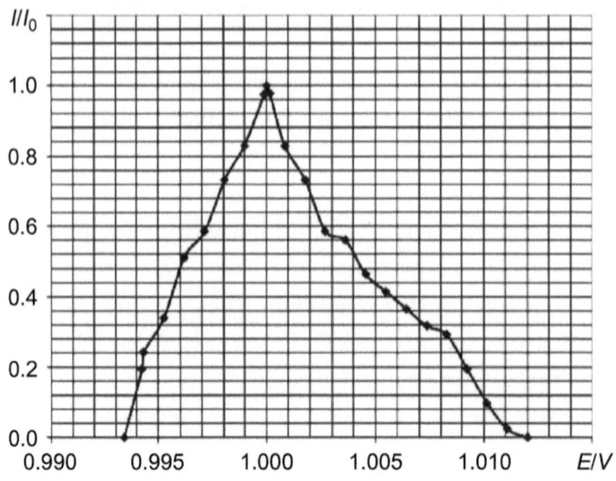

Figure 26 Instrumental function of the conical analyzer. (See the color plate.)

9.3 Systems Based on Spherical Optics

The electron-optical features of the electrostatic spherical mirror and relevant devices make them exceedingly promising in the analysis of photoelectron flows. A unique feature of electron mirror—namely, its capability for ideal focusing of electrons—was discovered by Sar-El (1966). It was shown that, for the reflection parameter $S = 1$, a point source is imaged into the diametrically opposite point without spherical aberration. Here,

$$S = \frac{qV}{2E} \frac{1}{1 - R_{in}/R_{out}}, \tag{96}$$

where q and E are the charge and kinetic energy of a charged particle, V is the deflection voltage applied between the outer electrode with the radius R_{out} and the inner electrode with the radius R_{in} (Figure 27). However, Sar-El (1966) limited its scope to the analysis of electron-optical systems with a close-to-zero linear energy dispersion, which, in addition, depends on the particle's input angle.

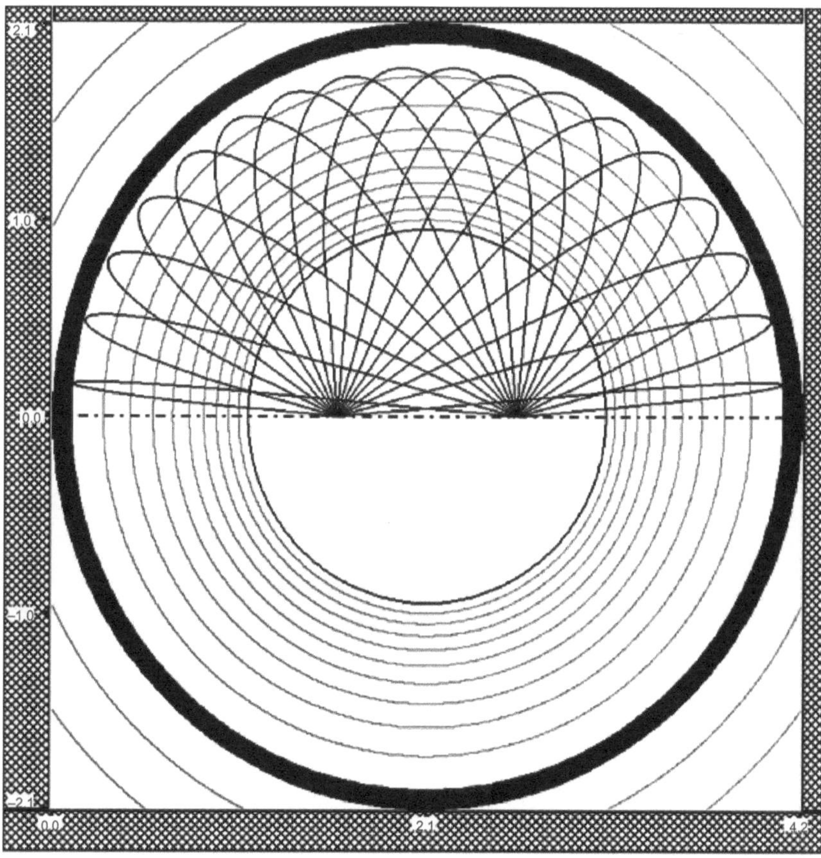

Figure 27 Charged particle trajectories in a spherical mirror. (See the color plate.)

A more complete study on electron-optical characteristics of spherical mirrors was performed by Zashkvara et al. (1988), who proposed a scheme in which the dispersion by energy does not depend on the input angle and equals the double value of the inner radius of a spherical electrode. In that scheme, the reflection parameter equals the unit ($S = 1$), while the source and the image are located at diametrically opposite points of the spherical inner electrode.

A promising and workable design of the energy analyzer developed by the authors of this chapter incorporates two semi-spheres with an output aperture located at the inner sphere's center, a fine-structure grid for electron beam transmission, and several correcting rings designated for fringe field compensation (Figure 28). The luminosity of such a device

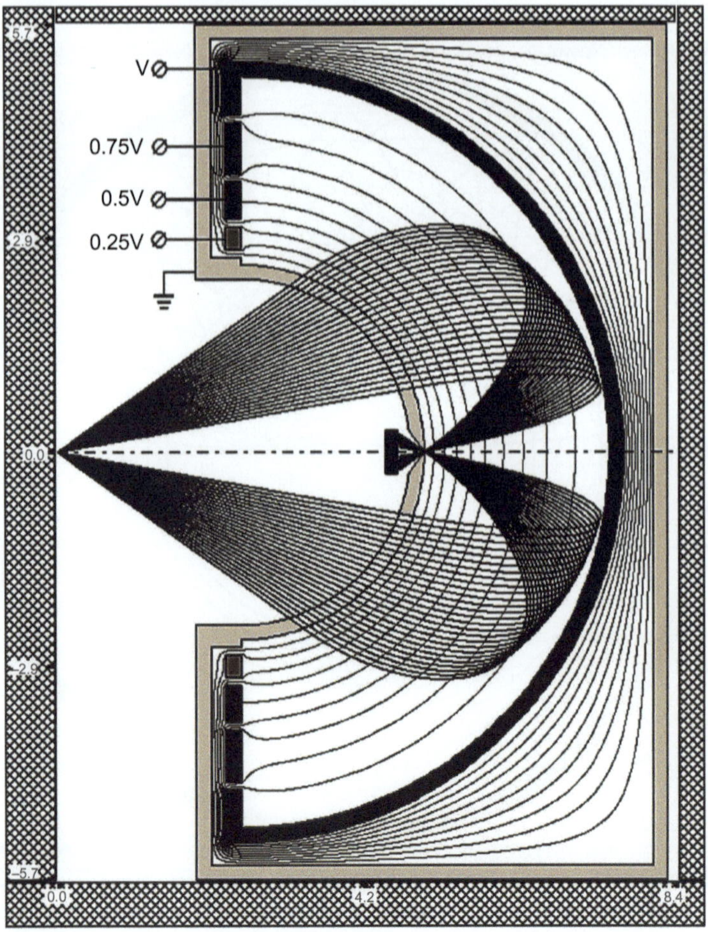

Figure 28 Actual construction of the spherical mirror. (See the color plate.)

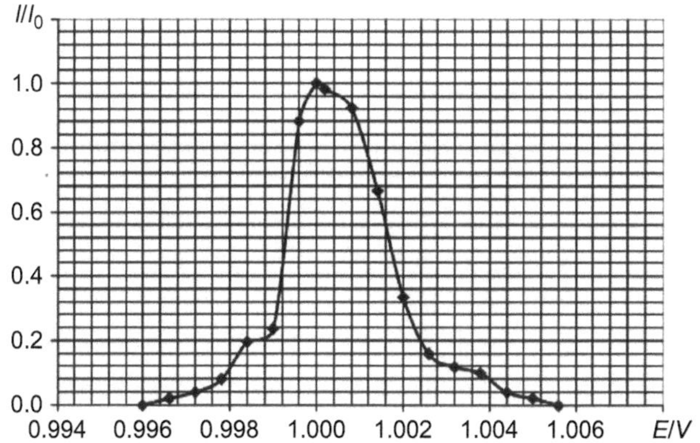

Figure 29 Instrumental function of the spherical mirror. (See the color plate.)

considerably exceeds that of all current designs, with the energy resolution being theoretically unrestricted from below.

Figure 29 shows the instrumental function of the spherical mirror with a range of initial angles α from 10° to 35°. The electron-optical scheme of the mirror is shown in Figure 28. The relative energy resolution at FWHM of the analyzer instrumental function with an output aperture radius of $0.00104R_{in}$ amounts to 0.24% at the luminosity $\Omega/2\pi = 16.6\%$.

APPENDICES

APPENDIX 1. SOME INTERMEDIATE MATHEMATICAL CALCULATIONS RELEVANT TO CHAPTER 3

This appendix contains some intermediate mathematical calculations made in the course of derivation of the recurrent relation to determine the coefficients of the power series [see Eq. (21) in Chapter 3].

Note: if an equation does not have "A" in its number, it is from Chapter 3.

A consistent calculation of terms in Eq. (17) gives

$$\frac{d\xi}{dx} = \frac{1}{2\sqrt{x}}\sum_{n=0}^{\infty} C_n x^n + \sqrt{x}\sum_{n=1}^{\infty} nC_n x^{n-1}, \tag{A1.1}$$

$$\left(\frac{d\xi}{dx}\right)^2 = \frac{1}{4x}\left(\sum_{n=0}^{\infty} C_n x^n\right)^2 + \left(\sum_{n=0}^{\infty} C_n x^n\right)\left(\sum_{n=1}^{\infty} nC_n x^{n-1}\right) + \left(\sum_{n=1}^{\infty} nC_n x^{n-1}\right)^2. \tag{A1.2}$$

It is convenient to determine the value $4x\left(\frac{d\xi}{dx}\right)^2$ first:

$$4x\left(\frac{d\xi}{dx}\right)^2 = C_0{}^2 + 6C_0C_1x + \left(9C_1^2 + 10C_0C_2\right)x^2 + (14C_0C_3 + 30C_1C_2)x^3$$
$$+ \left(25C_2^2 + 18C_0C_4 + 42C_1C_3\right)x^4 + (22C_0C_5 + 54C_1C_4 + 70C_2C_3)x^5$$
$$+ \left(26C_0C_6 + 66C_1C_5 + 90C_2C_4 + 49C_3^2\right)x^6\ldots, \tag{A1.3}$$

$$4x\left(\frac{d\xi}{dx}\right)^2 = h_0 + h_1x + h_2x^2 + h_3x^3 + h_4x^4 + h_5x^5 + h_6x^6\ldots. \tag{A1.4}$$

Let us represent Eq. (17) in the form

$$4x\left(\frac{d\xi}{dx}\right)^2\left[g(0) - g(x) - \int_0^x \varepsilon_2\xi'\,dx\right] = 4x\left[P^2\text{ctg}^2\alpha_0 - \int_x^{\rho_m} \varepsilon_2\xi'\,dx\right] \tag{A1.5}$$

and formulate some expansions that are necessary for further calculations:

$$\xi^2 = x\left(\sum_{n=0}^{\infty} C_n x^n\right)^2 = C_0^2 x + 2C_0C_1x^2 + \left(C_1^2 + 2C_0C_2\right)x^3 + 2(C_0C_3 + C_1C_2)x^4$$
$$+ \left(C_2^2 + 2C_0C_4 + 2C_1C_3\right)x^5 + 2(C_0C_5 + C_1C_4 + C_2C_3)x^6 \ldots, \tag{A1.6}$$

$$\rho = \rho_m - x,\ \rho + \frac{\rho^2}{2} = \rho_m\left(1 + \frac{\rho_m}{2}\right) - (1 + \rho_m)x + \frac{x^2}{2}. \tag{A1.7}$$

$$\ln(1+\rho) = \ln(1+\rho_m) + \ln\left(1 - \frac{x}{1+\rho_m}\right) = \ln(1+\rho_m) - \frac{x}{1+\rho_m} - \frac{x^2}{2(1+\rho_m)^2}$$
$$- \frac{x^3}{3(1+\rho_m)^3} - \frac{x^4}{4(1+\rho_m)^4} - \frac{x^5}{5(1+\rho_m)^5} - \frac{x^6}{6(1+\rho_m)^6} \cdots \tag{A1.8}$$

Using Eqs. (A1.6) and (A1.7), we can transform the expression in square brackets in Eq. (20) to the form

$$\xi^2 - \left(1 - \frac{2}{\gamma}\right) - \rho - \frac{\rho^2}{2} = -\left[1 - \frac{2}{\gamma} + \rho_m\left(1 + \frac{\rho_m}{2}\right)\right] + \left(C_0^2 + 1 + \rho_m\right)x$$
$$+ \left(2C_0C_1 - \frac{1}{2}\right)x^2 + \left(C_1^2 + 2C_0C_2\right)x^3 + 2(C_0C_3 + C_1C_2)x^4 \tag{A1.9}$$
$$+ 2\left(C_0C_4 + C_1C_3 + \frac{C_2^2}{2}\right)x^5 + 2(C_0C_5 + C_1C_4 + C_2C_3)x^6 \ldots.$$

By substituting the expansions (A1.7), (A1.8), and (A1.9) into Eq.(20), we obtain

$$g(x) = \frac{\gamma}{2}\{-\ln(1+\rho_m)[1 - 2/\gamma + \rho_m(1+\rho_m/2)] + \rho_m(1+\rho_m/2)$$
$$+ \left[\left(C_0^2 + 1 + \rho_m\right)\ln(1+\rho_m) + \frac{1}{1+\rho_m}(1 - 2/\gamma + \rho_m(1+\rho_m/2)) - 1 - \rho_m\right]x$$
$$+ \left[\begin{array}{l}\left(2C_0C_1 - \frac{1}{2}\right)\ln(1+\rho_m) - \left(C_0^2 + 1 + \rho_m\right)\frac{1}{1+\rho_m} + \\ + \frac{1}{2(1+\rho_m)^2}(1 - 2/\gamma + \rho_m(1+\rho_m/2)) + \frac{1}{2}\end{array}\right]x^2$$
$$+ \left[\begin{array}{l}\left(2C_0C_2 + C_1^2\right)\ln(1+\rho_m) - \frac{1}{1+\rho_m}\left(2C_0C_1 - \frac{1}{2}\right) - \\ - \frac{1}{2(1+\rho_m)^2}\left(C_0^2 + 1 + \rho_m\right) + \frac{1}{3(1+\rho_m)^3}(1 - 2/\gamma + \rho_m(1+\rho_m/2))\end{array}\right]x^3$$

$$+\left[\begin{array}{l}2(C_0C_3+C_1C_2)\ln(1+\rho_m)-\frac{1}{1+\rho_m}(2C_0C_2+C_1^2)-\frac{C_0C_1-1/4}{(1+\rho_m)^2}-\\-\frac{1}{3(1+\rho_m)^3}(C_0^2+1+\rho_m)+\frac{1}{4(1+\rho_m)^4}(1-2/\gamma+\rho_m(1+\rho_m/2))\end{array}\right]x^4$$

$$+\Big[2(C_0C_4+C_1C_3+C_2^2/2)\ln(1+\rho_m)-\frac{2}{1+\rho_m}(C_0C_3+C_1C_2)$$
$$-\frac{1}{2(1+\rho_m)^2}(2C_0C_2+C_2^2)-\frac{1}{3(1+\rho_m)^3}(2C_0C_1-1/2)$$
$$-\frac{1}{4(1+\rho_m)^4}(C_0^2+1+\rho_m)+\frac{1}{5(1+\rho_m)^5}(1-2/\gamma+\rho_m(1+\rho_m/2))]x^5$$
$$+\Big[2(C_0C_5+C_1C_4+C_2C_3)\ln(1+\rho_m)-\frac{2}{1+\rho_m}\left(C_0C_4+C_1C_3+\frac{C_2^2}{2}\right)$$
$$-\frac{1}{(1+\rho_m)^2}(C_0C_3+C_1C_2)-\frac{1}{3(1+\rho_m)^3}(2C_0C_2+C_1^2)$$
$$-\frac{1}{4(1+\rho_m)^4}(2C_0C_1-1/2)-\frac{1}{5(1+\rho_m)^5}(C_0^2+1+\rho_m)$$
$$+\frac{1}{6(1+\rho_m)^6}(1-2/\gamma+\rho_m(1+\rho_m/2))\Big]x^6+\ldots\}. \tag{A1.10}$$

$$g(x)=\frac{\gamma}{2}\{e_0+e_1x+e_2x^2+e_3x^3+e_4x^4+e_5x^5+e_6x^6+\ldots\}. \tag{A1.11}$$

Then

$$g(0)-g(x)=\frac{\gamma}{2}\{-e_1x-e_2x^2-e_3x^3-e_4x^4-e_5x^5-e_6x^6-\ldots\}. \tag{A1.12}$$

Let us return to Eq. (A1.5) and define the integrand $\varepsilon_2\xi'$ as a power series with respect to x. According to Eq. (20),

$$\varepsilon_2=-\frac{\partial g}{\partial \xi}=-\gamma\xi\ln(1+\rho),$$

so we have

$$\varepsilon_2\xi'=-\gamma\xi\xi'\ln(1+\rho). \tag{A1.13}$$

It follows from Eqs. (21) and (A1.1) that

$$\xi\xi'=\frac{1}{2}\left(\sum_{n=0}^{\infty}C_nx^n\right)^2+x\left(\sum_{n=0}^{\infty}C_nx^n\right)\left(\sum_{n=1}^{\infty}nC_nx^{n-1}\right), \tag{A1.14}$$

or

$$\begin{aligned}\xi\xi' &= C_0^2/2 + 2C_0C_1x + 3\left(C_0C_2 + C_1^2/2\right)x^2 + 4\left(C_0C_3 + C_1C_2\right)x^3 \\ &\quad + 5\left(C_0C_4 + C_1C_3 + C_2^2/2\right)x^4 + 6\left(C_0C_5 + C_1C_4 + C_2C_3\right)x^5 \\ &\quad + 7\left(C_0C_6 + C_1C_5 + C_2C_4 + C_3^2/2\right)x^6 \ldots.\end{aligned} \tag{A1.15}$$

Accordingly, from Eqs. (A1.8) and (A1.15), we get

$$\begin{aligned}\xi\xi' \ln(1+\rho) &= \frac{1}{2}C_0^2 \ln(1+\rho_m) + \left[2C_0C_1 \ln(1+\rho_m) - \frac{C_0^2}{2(1+\rho_m)}\right]x \\ &\quad + \left[3\left(C_0C_2 + C_1^2/2\right) \ln(1+\rho_m) - \frac{2C_0C_1}{1+\rho_m} - \frac{C_0^2}{4(1+\rho_m)^2}\right]x^2 \\ &\quad + \left[4(C_0C_3 + C_1C_2) \ln(1+\rho_m)\right. \\ &\quad \left. - \frac{3\left(C_0C_2 + C_1^2/2\right)}{1+\rho_m} - \frac{C_0C_1}{(1+\rho_m)^2} - \frac{C_0{}^2}{6(1+\rho_m)^3}\right]x^3 \\ &\quad + \left[5\left(C_0C_4 + C_3C_1 + C_2^2/2\right) \ln(1+\rho_m)\right. \\ &\quad \left. - \frac{4(C_0C_3 + C_1C_2)}{1+\rho_m} - \frac{3\left(C_0C_2 + C_1^2/2\right)}{2(1+\rho_m)^2} - \frac{2C_0C_1}{3(1+\rho_m)^3} - \frac{1}{8(1+\rho_m)^4}C_0^2\right]x^4 \\ &\quad + \left[6(C_0C_5 + C_1C_4 + C_2C_3) \ln(1+\rho_m) - \frac{5}{1+\rho_m}\left(C_0C_4 + C_1C_3 + C_2^2/2\right)\right. \\ &\quad \left. - \frac{2(C_0C_3 + C_1C_2)}{(1+\rho_m)^2} - \frac{C_0C_2 + C_1^2/2}{(1+\rho_m)^3} - \frac{C_0C_1}{2(1+\rho_m)^4} - \frac{1}{10(1+\rho_m)^5}C_0^2\right]x^5 \\ &\quad + \left[7\left(C_0C_6 + C_1C_5 + C_2C_4 + C_3^2/2\right) \ln(1+\rho_m)\right. \\ &\quad - \frac{6(C_0C_5 + C_1C_4 + C_2C_3)}{1+\rho_m} \\ &\quad - \frac{5\left(C_0C_4 + C_1C_3 + C_2^2/2\right)}{2(1+\rho_m)^2} - \frac{4(C_0C_3 + C_1C_2)}{3(1+\rho_m)^3} - \frac{3\left(C_0C_2 + C_1^2/2\right)}{4(1+\rho_m)^4} \\ &\quad \left. - \frac{2C_0C_1}{5(1+\rho_m)^5} - \frac{1}{12(1+\rho_m)^6}C_0^2\right]x^6 \ldots,\end{aligned} \tag{A1.16}$$

$$\xi\xi' \ln(1+\rho_m) = b_1 + b_2x + b_3x^2 + b_4x^3 + b_5x^4 + b_6x^5 + b_7x^6 \ldots. \tag{A1.17}$$

Now the sought-for integral can be represented in the form

$$\int_0^x \varepsilon_2 \xi' dx = -\gamma \int_0^x \xi\xi' \ln(1+\rho_m) dx$$

$$= -\gamma \left\{ b_1 x + \frac{b_2 x^2}{2} + \frac{b_3 x^3}{3} + \frac{b_4 x^4}{4} + \frac{b_5 x^5}{5} + \frac{b_6 x^6}{6} + \ldots \right\}. \qquad \text{(A1.18)}$$

For the sake of convenience, let us denote the expression in braces in Eq. (A1.18) as $\phi(x)$, then

$$\int_0^x \varepsilon_2 \xi' dx = -\gamma \phi(x). \qquad \text{(A1.19)}$$

In accordance with Eqs. (A1.12) and (A1.18), the equation in square brackets on the left side of Eq. (A1.5) appears as

$$g(0) - g(x) - \int_0^x \varepsilon_2 \xi' dx = \gamma \left[(b_1 - e_1/2)x + (b_2/2 - e_2/2)x^2 + (b_3/3 - e_3/2)x^3 \right.$$
$$+ (b_4/4 - e_4/2)x^4 + (b_5/5 - e_5/2)x^5 + (b_6/6 - e_6/2)x^6 \ldots. \qquad \text{(A1.20)}$$

Now, using Eqs. (A1.4) and (A1.20), let us represent the left side of Eq. (A1.5) in the form

$$4x\left(\frac{d\xi}{dx}\right)^2 \left[g(0) - g(x) - \int_0^x \varepsilon_2 \xi' dx \right] = \gamma \{ h_0(b_1 - e_1/2)x + [h_0(b_2/2 - e_2/2)$$
$$+ h_1(b_1 - e_1/2)]x^2 + [h_0(b_3/3 - e_3/2) + h_1(b_2/2 - e_2/2) + h_2(b_1 - e_1/2)]x^3$$
$$+ [h_0(b_4/4 - e_4/2) + h_1(b_3/3 - e_3/2) + h_2(b_2/2 - e_2/2) + h_3(b_1 - e_1/2)]x^4 \qquad \text{(A1.21)}$$
$$+ [h_0(b_5/5 - e_5/2) + h_1(b_4/4 - e_4/2) + h_2(b_3/3 - e_3/2) + h_3(b_2/2 - e_2/2)$$
$$+ h_4(b_1 - e_1/2)]x^5 + [h_0(b_6/6 - e_6/2) + h_1(b_5/5 - e_5/2) + h_2(b_4/4 - e_4/2)$$
$$+ h_3(b_3/3 - e_3/2) + h_4(b_2/2 - e_2/2) + h_5(b_1 - e_1/2)]x^6 + \ldots \}.$$

Calculation of the integral on the right side of Eq.(A1.5) gives

$$\int_x^{\rho_m} \varepsilon_2 \xi' dx = -\gamma \left\{ b_1\rho_m + \frac{b_2\rho_m{}^2}{2} + \frac{b_3\rho_m{}^3}{3} + \frac{b_4\rho_m{}^4}{4} + \frac{b_5\rho_m{}^5}{5} + \frac{b_6\rho_m{}^6}{6} + \ldots \right\}$$
$$+\gamma \left\{ b_1 x + \frac{b_2 x^2}{2} + \frac{b_3 x^3}{3} + \frac{b_4 x^4}{4} + \frac{b_5 x^5}{5} + \frac{b_6 x^6}{6} + \ldots \right\},$$
$$\int_x^{\rho_m} \varepsilon_2 \xi' dx = \gamma \left\{ -F(\rho_m) + b_1 x + \frac{b_2 x^2}{2} + \frac{b_3 x^3}{3} + \frac{b_4 x^4}{4} + \frac{b_5 x^5}{5} + \frac{b_6 x^6}{6} + \ldots \right\}, \tag{A1.22}$$

where

$$F(\rho_m) = b_1\rho_m + \frac{b_2\rho_m{}^2}{2} + \frac{b_3\rho_m{}^3}{3} + \frac{b_4\rho_m{}^4}{4} + \frac{b_5\rho_m{}^5}{5} + \frac{b_6\rho_m{}^6}{6} + \ldots. \tag{A1.23}$$

Now we are able to represent the right side of Eq. (A1.5) as

$$4x\left[P^2 \mathrm{ctg}^2\alpha_0 - \int_x^{\rho_m} \varepsilon_2 \xi' dx\right] = 4\left[P^2 \mathrm{ctg}^2\alpha_0 + \gamma F(\rho_m)\right]x$$
$$-4\gamma b_1 x^2 - 2\gamma b_2 x^3 - \frac{4\gamma b_3 x^4}{3} - \gamma b_4 x^5 - \frac{4\gamma b_5 x^6}{5} - \ldots. \tag{A1.24}$$

Thus, both sides of Eq. (A1.5) are represented in the form of power series. By equating the terms at equal power in Eqs. (A1.21) and (A1.24), we can determine the coefficients C_n of the series [Eq. (21)], which allows us to analyze the electron-optical characteristics of the system in question. In doing so, we obtain the following system of equations:

$$\gamma h_0(b_1 - e_1/2) = 4\left[P^2 \mathrm{ctg}^2\alpha_0 + \gamma F(\rho_m)\right],$$
$$h_0(b_2/2 - e_2/2) + h_1(b_1 - e_1/2) = 4b_1,\ h_0(b_3/3 - e_3/2) + h_1(b_2/2 - e_2/2)$$
$$+ h_2(b_1 - e_1/2) = -2b_2,$$
$$h_0(b_4/4 - e_4/2) + h_1(b_3/3 - e_3/2) + h_2(b_2/2 - e_2/2) + h_3(b_1 - e_1/2)$$
$$= -4b_3/3, \tag{A1.25}$$
$$h_0(b_5/5 - e_5/2) + h_1(b_4/4 - e_4/2) + h_2(b_3/3 - e_3/2)$$
$$+ h_3(b_2/2 - e_2/2) + h_4(b_1 - e_1/2) = -b_4,$$

$$h_0(b_6/6-e_6/2)+h_1(b_5/5-e_5/2)+h_2(b_4/4-e_4/2)+h_3(b_3/3-e_3/2)$$
$$+h_4(b_2/2-e_2/2)+h_5(b_1-e_1/2)=-4b_5/5.$$

Using the expressions obtained for the coefficients h_i, e_i, and b_i, we get the following equations:

$$\begin{aligned}
h_0 &= C_0^2 \\
h_1 &= 6C_0C_1 \\
h_2 &= 9C_1^2+10C_0C_2 \\
h_3 &= 14C_0C_3+30C_1C_2 \\
h_4 &= 25C_2^2+18C_0C_4+42C_1C_3 \\
h_5 &= 22C_0C_5+54C_1C_4+70C_2C_3 \\
h_6 &= 26C_0C_6+66C_1C_5+90C_2C_4+49C_3^2
\end{aligned} \quad (A1.26)$$

$$e_0=\rho_m(1+\rho_m/2)-[1-2/\gamma+\rho_m(1+\rho_m/2)]\ln(1+\rho_m)$$

$$e_1=\left(1+\rho_m+C_0^2\right)\ln(1+\rho_m)+[1-2/\gamma+\rho_m(1+\rho_m/2)]/(1+\rho_m)-1-\rho_m$$

$$e_2=(2C_0C_1-1/2)\ln(1+\rho_m)-\left(1+\rho_m+C_0^2\right)/(1+\rho_m)+\frac{1}{2(1+\rho_m)^2}[1-2/\gamma+\rho_m(1+\rho_m/2)]+\frac{1}{2}$$

$$e_3=\left(2C_0C_2+C_1^2\right)\ln(1+\rho_m)-(2C_0C_1-1/2)/(1+\rho_m)-\frac{1}{2(1+\rho_m)^2}\left(1+\rho_m+C_0^2\right)+\frac{1}{3(1+\rho_m)^3}[1-2/\gamma+\rho_m(1+\rho_m/2)]$$

$$e_4=2(C_0C_3+C_1C_2)\ln(1+\rho_m)-\frac{2C_0C_2+C_1^2}{1+\rho_m}-(C_0C_1-1/4)/(1+\rho_m)^2-\frac{1}{3(1+\rho_m)^3}\left(1+\rho_m+C_0^2\right)+\frac{1}{4(1+\rho_m)^4}[1-2/\gamma+\rho_m(1+\rho_m/2)]$$

$$(A1.27)$$

$$e_5=2\left(C_0C_4+C_1C_3+C_2^2/2\right)\ln(1+\rho_m)-\frac{2}{1+\rho_m}(C_0C_3+C_1C_2)-\frac{1}{2(1+\rho_m)^2}\left(2C_0C_2+C_1^2\right)-\frac{1}{3(1+\rho_m)^3}(2C_0C_1-1/2)-\frac{1}{4(1+\rho_m)^4}\left(1+\rho_m+C_0^2\right)+\frac{1}{5(1+\rho_m)^5}[1-2/\gamma+\rho_m(1+\rho_m/2)]$$

$$e_6 = 2(C_0C_5 + C_1C_4 + C_2C_3)\ \ln(1+\rho_m) - \frac{2}{1+\rho_m}\left(C_0C_4 + C_1C_3 + C_2^2/2\right)$$
$$- \frac{1}{(1+\rho_m)^2}(C_0C_3 + C_1C_2) - \frac{1}{3(1+\rho_m)^3}\left(2C_0C_2 + C_1^2\right) - \frac{2C_0C_1 - 1/2}{4(1+\rho_m)^4}$$
$$- \frac{1}{5(1+\rho_m)^5}\left(1+\rho_m + C_0^2\right) + \frac{1}{6(1+\rho_m)^6}[1 - 2/\gamma + \rho_m(1+\rho_m/2)]$$
$$b_1 = \frac{1}{2}C_0^2\ \ln(1+\rho_m)$$
$$b_2 = 2C_0C_1\ \ln(1+\rho_m) - \frac{1}{2}\frac{C_0^2}{1+\rho_m}$$
$$b_3 = 3\left(C_0C_2 + C_1^2/2\right)\ \ln(1+\rho_m) - 2C_0C_1/(1+\rho_m) - \frac{C_0^2}{4(1+\rho_m)^2}$$
$$b_4 = 4(C_0C_3 + C_1C_2)\ \ln(1+\rho_m) - \frac{3\left(C_0C_2 + C_1^2/2\right)}{1+\rho_m} - \frac{C_0C_1}{(1+\rho_m)^2}$$
$$- \frac{C_0^2}{6(1+\rho_m)^3}$$
$$b_5 = 5\left(C_0C_4 + C_1C_3 + C_2^2/2\right)\ \ln(1+\rho_m) - 4(C_0C_3 + C_1C_2)/(1+\rho_m)$$
$$- \frac{3}{2}\frac{C_0C_2 + C_1^2/2}{(1+\rho_m)^2} - \frac{2}{3}\frac{C_0C_1}{(1+\rho_m)^3} - \frac{C_0^2}{8(1+\rho_m)^4}$$
$$b_6 = 6(C_0C_5 + C_1C_4 + C_2C_3)\ \ln(1+\rho_m) - 5\left(C_0C_4 + C_1C_3 + C_2^2/2\right)/(1+\rho_m)$$
$$- \frac{2(C_0C_3 + C_1C_2)}{(1+\rho_m)^2} - \frac{C_0C_2 + C_1^2/2}{(1+\rho_m)^3} - \frac{C_0C_1}{2(1+\rho_m)^4} - \frac{C_0^2}{10(1+\rho_m)^5} \tag{A1.28}$$
$$b_7 = 7\left(C_0C_6 + C_1C_5 + C_2C_4 + C_3^2/2\right)\ \ln(1+\rho_m)$$
$$-6(C_0C_5 + C_1C_4 + C_2C_3)/(1+\rho_m) - \frac{5\left(C_0C_4 + C_1C_3 + C_2^2/2\right)}{2(1+\rho_m)^2}$$
$$- \frac{4(C_0C_3 + C_1C_2)}{3(1+\rho_m)^3} - \frac{3\left(C_0C_2 + C_1^2/2\right)}{4(1+\rho_m)^4} - \frac{2C_0C_1}{5(1+\rho_m)^5} - \frac{C_0^2}{12(1+\rho_m)^6}$$

APPENDIX 2. THE RELATIONSHIP BETWEEN THE OUTPUT PARAMETERS OF THE ENERGY ANALYZERS CONSIDERED IN CHAPTER 3 AND INITIAL AND BOUNDARY CONDITIONS IN DIFFERENT MODES

In some practical cases, especially when the energy analyzer represents only part of an entire experimental installation, the choice can be made in favor of the analyzer that does not possess the highest possible electron-optical performance. Tables A2.1 through A2.3 of this appendix provide a fairly wide range of the calculated modes that can be useful in practical implementation.

Table A2.1 Electron-Optical Characteristics of the Hexapole-Cylindrical Energy Analyzer with End-Face Electrodes

P	α_o	ρ_m	$\Delta=\Delta_1+\Delta_2$	l	D	A_{lll}	$\|\delta\|$
0.100	30.3220	0.0101	0.0202	0.1033	0.0693	−0.3124	652.23
0.150	30.7035	0.0228	0.0558	0.2311	0.1565	−0.6871	669.42
0.200	31.1971	0.0410	0.0828	0.4089	0.2806	−1.1994	687.69
0.250	31.7664	0.0652	0.1332	0.6386	0.4465	−1.8735	700.42
0.275	32.0648	0.0799	0.1644	0.7747	0.5481	−2.8027	701.68
0.300	32.3623	0.0964	0.2004	0.9271	0.6650	−2.1982	697.32
0.325	32.6495	0.1151	0.2424	1.0985	0.8007	−3.4336	685.34
0.350	32.9138	0.1364	0.2920	1.2929	0.9605	−4.2554	663.35
0.375	33.1392	0.1607	0.3516	1.5170	1.1528	−5.3871	628.93
0.400	33.3039	0.1889	0.4254	1.7819	1.3917	−7.0528	579.92
0.425	33.3778	0.2224	0.5201	2.1066	1.7014	−9.7085	515.05
0.450	33.3175	0.2634	0.6492	2.5264	2.1280	−14.3886	434.64
0.475	33.0537	0.3165	0.8406	3.1154	2.7679	−23.8499	341.07
0.489	32.7670	0.3557	1.00	3.5870	3.3065	−34.2724	283.53
0.500	32.4517	0.3925	1.1643	4.0613	3.8644	−47.6204	238.49
0.525	31.1491	0.5257	1.8790	6.0619	6.3010	−139.9162	132.35
0.527	30.9524	0.5453	2.00	6.3964	6.7137	−161.4419	122.22

Table A2.2 Electron-Optical Characteristics of the Hexapole-Cylindrical Analyzer with $\gamma = 1$

P	α_0 (°)	ρ_m	Δ	ξ_m	l	D	A_{lll}
0.10	30.3300	0.0100	0.0101	0.0344	0.1032	0.0693	−0.3114
0.12	30.4780	0.0145	0.0145	0.0494	0.1481	0.0997	−0.4426
0.14	30.6510	0.0198	0.0198	0.0670	0.2009	0.1356	−0.5932
0.16	30.8520	0.0259	0.0260	0.0871	0.2611	0.1769	−0.7604
0.18	31.0790	0.0328	0.0330	0.1097	0.3287	0.2235	−0.9417
0.20	31.3342	0.0406	0.0408	0.1347	0.4033	0.2752	−1.1334
0.22	31.6161	0.0493	0.0495	0.1619	0.4845	0.3320	−1.3320
0.24	31.9253	0.0588	0.0590	0.1913	0.5720	0.3936	−1.5331
0.26	32.2613	0.0691	0.0694	0.2227	0.6651	0.4596	−1.7324
0.28	32.6238	0.0803	0.0805	0.2560	0.7636	0.5298	−1.9254
0.30	33.0120	0.0922	0.0924	0.2911	0.8667	0.6037	−2.1075
0.32	33.4251	0.1050	0.1050	0.3278	0.9740	0.6809	−2.2744
0.34	33.8615	0.1185	0.1184	0.3660	1.0848	0.7609	−2.4222
0.36	34.3192	0.1328	0.1323	0.4055	1.1987	0.8431	−2.5475
0.38	34.7954	0.1477	0.1468	0.4463	1.3151	0.9272	−2.6471
0.40	35.2868	0.1633	0.1618	0.4882	1.4337	1.0125	−2.7177
0.42	35.7893	0.1795	0.1773	0.5311	1.5541	1.0987	−2.7551
0.44	36.2988	0.1963	0.1932	0.5750	1.6761	1.1852	−2.7533
0.46	36.8114	0.2136	0.2094	0.6199	1.7994	1.2720	−2.7026
0.48	37.3254	0.2312	0.2259	0.6656	1.9237	1.3588	−2.5876
0.50	37.8455	0.2493	0.2425	0.7120	2.0482	1.4457	−2.3843
0.52	38.3918	0.2679	0.2588	0.7586	2.1704	1.5330	−2.0545
0.54	39.0245	0.2869	0.2741	0.8038	2.2841	1.6199	−1.5385
0.56	39.9046	0.3066	0.2861	0.8439	2.3721	1.7016	−0.7604
0.5754	40.9766	0.3225	0.2899	0.8657	2.3991	1.7516	0.0000
0.58	41,3980	0.3274	0.2896	0.8698	2.3965	1.7626	0.2201
0.60	43.7420	0.3498	0.2811	0.8763	2.3401	1.7967	0.7340

Table A2.2 Electron-Optical Characteristics of the Hexapole-Cylindrical Analyzer with $\gamma = 1$—cont'd

P	α_0 (°)	ρ_m	Δ	ξ_m	I	D	A_{III}
0.62	46.0910	0.3722	0.2736	0.8840	2.2948	1.8435	0.2135
0.6242	46.5065	0.3766	0.2734	0.8873	2.2934	1.8568	0.0000
0.64	47.8170	0.3922	0.2779	0.9063	2.3162	1.9178	−1.0781
0.66	48.7420	0.4071	0.2978	0.9468	2.4161	2.0175	−3.0426
0.68	47.9590	0.4121	0.3473	1.0273	2.6809	2.2024	−6.9744
0.70	45.0826	0.4263	0.4703	1.1806	3.2990	2.9194	−26.4767
0.7045	44.8600	0.4321	0.5000	1.2080	3.4207	3.1015	−33.1633
0.72	44.4548	0.4543	0.6199	1.2991	3.8617	3.7915	−65.2819

Table A2.3 Electron-Optical Characteristics of the Energy Analyzer with Potential Distribution $U(r,z) = \frac{5}{2}\ln r - U_h(r,z)$

P	α_0 (°)	ρ_m	Δ	ξ_m	I	D	A_{III}	δ
0.1	30.13140	0.0040	0.0040	0.0138	0.0415	0.0276	−0.1267	−641.14
0.2	30.51530	0.0161	0.0162	0.0549	0.1647	0.1105	−0.4926	−659.08
0.3	31.12230	0.0367	0.0370	0.1222	0.3670	0.2496	−1.0674	−687.16
0.4	31.90780	0.0664	0.0675	0.2151	0.6470	0.4493	−1.8347	−719.71
0.5	32.81590	0.1063	0.1097	0.3336	1.0074	0.7203	−2.8305	−747.84
0.6	33.77920	0.1585	0.1673	0.4801	1.4606	1.0841	−4.1958	−759.37
0.7	34.71893	0.2263	0.2476	0.6609	2.0365	1.5821	−6.2752	−740.96
0.8	35.54527	0.3154	0.3643	0.8888	2.7973	2.2910	−9.8689	−682.26
0.9	36.15655	0.4363	0.5458	1.1893	3.8723	3.3580	−17.0159	−579.97
1.0	36.43013	0.6094	0.8561	1.6154	5.5506	5.0859	−33.8878	−441.07

APPENDIX 3. BOUNDARY ELEMENT METHOD (BEM) FOR CALCULATING THE POTENTIAL AND ITS GRADIENT IN PLANAR SYSTEMS

Approximations by means of two-dimensional (2D) models are applicable to a variety of electron-optical systems, at least at the preliminary design stage. In this appendix, some specialized techniques are outlined, which ensure high-accuracy calculation of electrostatic potential and field distribution in planar electron-optical systems with electrodes of real thickness and virtually arbitrary geometry. The results of testing are also presented here.

A3.1 Using the Boundary Elements Method (BEM) to Solve the Planar Dirichlet Problem

The planar Dirichlet problem consists of finding the potential distribution $u = u(\xi)$, $\xi \in \Omega$ in the planar domain $\Omega(x,y)$, which fulfills the condition $u = u(\chi)$ on the boundary Γ ($\chi \in \Gamma$). We assume that the boundary constitutes a set of closed contours (electrodes of nonzero thickness), on each of which a constant potential is defined. Figure A3.1 demonstrates an example of the electron-optical system, the boundary of which represents two electrodes (contours 1 and 2), with potentials V_1 and V_2, respectively.

By neglecting the space-charge effects and using the second Green formula (Greenberg, 1971), the exterior planar Dirichlet problem can be reduced to the integral equation as follows:

$$\gamma(\xi)u(\xi) + \int_\Gamma u(\chi)Q^*(\xi,\chi)d\Gamma(\chi) = \int_\Gamma q(\chi)U^*(\xi,\chi)d\Gamma(\chi), \qquad \text{(A3.1)}$$

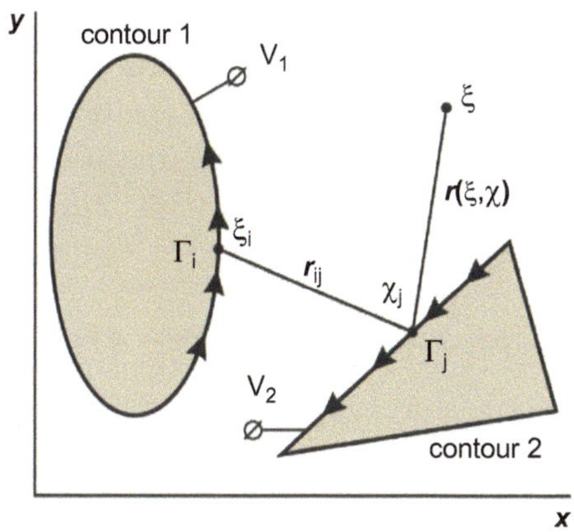

Figure A3.1 An example of representation of the boundary of a calculated domain by a set of closed contours and subdivision of the boundary into the boundary elements.

where the function $Q^*(\xi,\chi)$ is the normal derivative of the fundamental solution $U^*(\xi,\chi)$ [i.e., $Q^*(\xi,\chi)=\partial U^*/\partial n$ and $q(\chi)$ is the normal derivative of the boundary potential; i.e., $q(\chi)=du(\chi)/dn$, $\xi\in\Omega\cup\Gamma$]; the factor $\gamma(\xi)=2\pi$ for $\xi\in\Omega$ and $\gamma(\xi)$ is determined by the boundary configuration when $\xi\in\Gamma$. Here, n is the outward normal to Γ, and U^* is the fundamental solution of the Laplace equation.

The fundamental solution of Eq. (A3.1) in planar case is the natural logarithm of the distance between the points ξ and χ (Figure A3.1):

$$U^*(\xi,\chi)=-\ \ln r(\xi,\chi)=-\frac{1}{2}\ln\left([x(\chi)-x(\xi)]^2+[y(\chi)-y(\xi)]^2\right). \quad \text{(A3.2)}$$

The derivative of the fundamental solution $Q^*(\xi,\chi)=\frac{dU^*(\xi,\chi)}{dn}$ along the outward normal $\vec{n}\,(n_x,n_y)$ to the boundary Γ is defined by the following formulas:

$$\frac{dU^*(\xi,\chi)}{dn}=\operatorname{grad}U^*\cdot\vec{n}\,,\ \operatorname{grad}U^*=\frac{\partial U^*}{\partial x(\chi)}\vec{i}+\frac{\partial U^*}{\partial y(\chi)}\vec{j}\,,$$
$$\vec{n}=n_x\vec{i}+n_y\vec{j} \quad \text{(A3.3)}$$

where

$$\frac{\partial U^*}{\partial x}=-\frac{x(\chi)-x(\xi)}{r^2(\xi,\chi)},\ \frac{\partial U^*}{\partial y}=-\frac{y(\chi)-y(\xi)}{r^2(\xi,\chi)}. \quad \text{(A3.4)}$$

For the numerical solution of the Dirichlet problem, a discrete form of Eq. (A3.1) is formed by subdivision of the boundary Γ into N boundary elements Γ_j (Figure A3.1). Taking into consideration the constancy of the potential on each of the contours (electrodes), and assuming that the normal derivative of the potential is constant within each boundary element, Eq. (A3.1) can be written as

$$\gamma u(\xi) + \sum_{j=1}^{N} u_j H_j(\xi) = \sum_{j=1}^{N} q_j H_j(\xi). \tag{A3.5}$$

If the boundary Γ is approximated by the straight-line segments (vectors $\vec{\Gamma}_j$), the computation formulas for the functions $F_j(\xi)$ and $H_j(\xi)$ appear as follows:

$$F_j(\xi) = \Delta\chi_j \int_0^1 U^*\left(\xi, \Delta\chi_j \tau + \chi_j\right) d\tau = \Delta\chi_j \int_0^1 U_j^*(\xi, \tau) d\tau \tag{A3.6}$$

$$H_j(\xi) = \Delta\chi_j \int_0^1 Q^*\left(\xi, \Delta\chi_j \tau + \chi_j\right) d\tau = \Delta\chi_j \int_0^1 Q_j^*(\xi, \tau) d\tau, \tag{A3.7}$$

where the normalized integration variable $\tau \in [0,1]$ is introduced to simplify the calculations. Also in this equation, $\Delta\chi_j$ is the length of the rectilinear boundary element Γj, and χ_j is the initial point of the jth boundary element on the electrode contour. We will see next that these integrals possess analytical representations in the 2D case.

A3.2 Calculation of the Regular Integral $F_j(\xi)$

Using the normalized integration variable τ, the Cartesian coordinates of internal points of the jth boundary element can be expressed as

$$x_j(\tau) = a_j \tau + b_j, \; y_j(\tau) = c_j \tau + d_j, \tag{A3.8}$$

where $a_j = \cos(\varphi_j)\Delta\chi_j$, $c_i = \sin(\varphi_j)\Delta\chi_j$; and b_j and d_j are the coordinates of the initial point of the element Γ_j, φ_j is its angle relative to the x-axis. In this case,

$$\begin{aligned} U^*(\xi, \tau) &= -\frac{1}{2} \ln\left([x(\chi) - x(\xi)]^2 + [y(\chi) - y(\xi)]^2\right) \\ &= -\frac{1}{2} \ln\left[T_j \tau^2 - 2D_j(\xi)\tau + P_j(\xi)\right], \end{aligned} \tag{A3.9}$$

where $T_j = a_j^2 + c_j^2$, $D_j(\xi) = a_j\left[x(\xi) - b_j\right]$, $P_j(\xi) = \left[x(\xi) - b_j\right]^2 + \left[y(\xi) - d_j\right]^2$.

The integral over τ of the function $U_j^*(\xi, \tau)$ [see Eq. (A3.6)] is

$$F_j(\xi)=\Delta\chi_j\left\{\frac{1}{2}\left[\left(\frac{D_j(\xi)}{T_j}-1\right)\ln\left(T_j-2D_j(\xi)+P_j(\xi)\right)\right]\right.$$
$$\left.-\frac{D_j(\xi)}{T_j}\ln P_j(\xi)-\frac{Z_j(\xi)}{T_j}A_j(\xi)+1\right\},\tag{A3.10}$$

where $Z_j(\xi)=\sqrt{T_jP_j(\xi)-D_j^2(\xi)}$, $A_j(\xi)=\arctan\frac{D_j(\xi)}{Z_j(\xi)}+\arctan\frac{T_j-D_j(\xi)}{Z_j(\xi)}$.

A3.3 Calculation of the Regular Integral *H_j(ξ)*

Given Eqs. (A3.3) and (A3.8), the normal derivative of the fundamental solution can be written as

$$Q_j^*(\xi,\tau)=\frac{dU_j^*}{dn}$$
$$=-\frac{\left(a_jn_x+c_jn_y\right)\tau+\left[b_j-x(\xi)\right]n_x+\left[d_j-y(\xi)\right]n_y}{T_j\tau^2-2D_j(\xi)\tau+P_j(\xi)}.\tag{A3.11}$$

The sum $\left(a_jn_x+c_jn_y\right)$ equals zero as the scalar product of two mutually perpendicular vectors—namely, the boundary element $\vec{\Gamma}_j\left(a_j,c_j\right)$ and its unit normal vector $\vec{n}_j\left(n_x,n_y\right)$. Thus, integration of the normal derivative gives the following expression for the integral $H_j(\xi)$:

$$H_j(\xi)=\Delta\chi_jG_j(\xi)\int_0^1\frac{d\tau}{T_j\tau^2-2D_j(\xi)\tau+P_j(\xi)}$$
$$=\Delta\chi_j\frac{G_j(\xi)}{Z_j(\xi)}\left[\arctan\frac{D_j(\xi)}{Z_j(\xi)}+\arctan\frac{T_j-D_j(\xi)}{Z_j(\xi)}\right],\tag{A3.12}$$

where $G_j(\xi)=\left[x(\xi)-b_j\right]n_x+\left[y(\xi)-d_j\right]n_y$.

A3.4 Calculation of Potential Distribution

Calculation of electrostatic potential is performed in two stages. First, the unknown vector q_j for a given boundary potential distribution is calculated using Eq. (A3.5); i.e., the inverse problem is solved. Then the values q_j and the given values u_j are used to determine the function $u(\xi)$, $\xi\in\Omega$ from Eq. (A3.5); i.e., the direct problem is solved.

A3.4.1 Singularity Problem

To solve the inverse problem, the collocation technique is used. In accordance with this technique, nodes ξ_i are placed in the middle (i.e., $\xi_i=1/2$) of each rectilinear $[\gamma(\xi) = \pi]$ element Γ_i and system of the N equation [Eq. (A3.5)] for the whole set of N nodes ξ_i is written as

$$\pi u(\xi_i) + \Sigma_{j=1}^{N} u_j H_{ij} = \Sigma_{(j=1)}^{N} q_j F_{ij}, \qquad i = 1,2 \ldots N, \tag{A3.13}$$

where $H_{ij} = H_j(\xi_i)$, $F_{ij} = F_j(\xi_i)$.

We are faced here with a problem of singularity in the integrands, which is peculiar to the application of the integral technique to potential theory. If $i = j$, the singularity arises in the kernels of the integrals [Eqs. (A3.6) and (A3.7)]; i.e., when the integration is performed over the element Γ_j (this element is referred to as a "singular one") containing the current collocation node ξ_i.

A3.4.1.1 Calculation of H_{ii}

For both a singular planar boundary element in three-dimensional (3D) problems and a singular rectilinear element in 2D problems, the definition domain of the function $U^*(\xi,\chi)$ coincides with the plane of that element; therefore, $\operatorname{grad} U^*(\xi,\chi) \perp \vec{n}$, so that $Q^*(\xi,\chi) = \partial U^*/\partial n = \operatorname{grad} U^*(\xi,\chi) \cdot \vec{n} = 0$. Thus, $H_{ii} = 0$ in the interior Dirichlet problem. Due to the jump in the double-layer potential when crossing the boundary from the inside out (see Sternberg & Smith, 1944), we arrive at the following value of the integral H_{ii} for the exterior problem:

$$H_{ii} = 2\pi. \tag{A3.14}$$

A3.4.1.2 Calculation of F_{ii}

Given the fact that $\xi_i = 1/2$ and taking into consideration Eq. (A3.8) for the coordinates of the points of the ith boundary element, the singular fundamental solution can be transformed to the following form:

$$\begin{aligned} U_i^*(\xi_i, \tau) = U_{ii}^*(\tau) &= -\ln\sqrt{[x_i(\tau) - x(\xi_i)]^2 + [y_i(\tau) - y(\xi_i)]^2} \\ &= -\ln\sqrt{\left[a_i\left(\tau - \frac{1}{2}\right)\right]^2 + \left[c_i\left(\tau - \frac{1}{2}\right)\right]^2} \\ &= -\ln\left(|\tau - 1/2| \cdot \sqrt{T_i}\right). \end{aligned} \tag{A3.15}$$

This expression gives a final formula for the calculation of the integral [Eq. (A3.6)] upon the singular boundary element:

$$F_{ii} = -\Delta\chi_i \int_0^1 U_{ii}^*(\tau)d\tau = \Delta\chi_i\left(1 + \ln\frac{2}{\sqrt{T_i}}\right) \tag{A3.16}$$

A3.4.2 Inverse Problem Solution

In accordance with Eq. (A3.13), the inverse problem solution q_j can be obtained from the system of linear equations:

$$b_i = F_{ij}q_j, \tag{A3.17}$$

where $b_i = \pi u_i + \sum_{j=1}^{N} u_j H_{ij}$, $i, j = 1, 2 \ldots N$; integrals F_{ij} and H_{ij} are calculated from Eqs. (A3.10) and (A3.12) for $i \neq j$ and Eqs. (A3.16) and (A3.14) for $i = j$.

A3.4.3 Direct Problem Solution

The direct problem solution takes the following form [see Eq. (A3.5)]:

$$u(\xi) = \frac{1}{2\pi}\left[\sum_{j=1}^{N} u_j H_j(\xi) - \sum_{j=1}^{N} q_j F_j(\xi)\right], \xi \in \Omega. \tag{A3.18}$$

A3.5 Field Calculation

The potential gradient being an indispensable element of virtually all simulation problems in electron optics can be numerically defined using the standard formulas of numerical differentiation. These formulas have substantial restrictions on the calculation accuracy. In the case of planar electric field, the BEM allows us to derive certain analytic expressions in the form of finite series for the gradient components E_x and E_y [see Eq. (A3.18)]:

$$E_x = \frac{\partial u(\xi)}{\partial x} = \frac{1}{2\pi}\left[\sum_{j=1}^{N} q_j \frac{\partial F_j(\xi)}{\partial x} - \sum_{j=1}^{N} u_j \frac{\partial H_j(\xi)}{\partial x}\right] \tag{A3.19}$$

$$E_y = \frac{\partial u(\xi)}{\partial y} = \frac{1}{2\pi}\left[\sum_{j=1}^{N} q_j \frac{\partial F_j(\xi)}{\partial y} - \sum_{j=1}^{N} u_j \frac{\partial H_j(\xi)}{\partial y}\right]. \tag{A3.20}$$

The derivatives $\frac{\partial H_j(\xi)}{\partial x}$, $\frac{\partial H_j(\xi)}{\partial y}$, $\frac{\partial F_j(\xi)}{\partial x}$ and $\frac{\partial F_j(\xi)}{\partial y}$ appear as

$$\frac{\partial H_j(\xi)}{\partial x} = \frac{\Delta\chi_j}{Z_j^2(\xi)}\left\{\frac{n_x Z_j^2(\xi) - G_j(\xi)V_{xj}(\xi)}{Z_j(\xi)} A_j(\xi) + G_j(\xi)W_{xj}(\xi)\right\}, \tag{A3.21}$$

$$\frac{\partial H_j(\xi)}{\partial y} = \frac{\Delta\chi_j}{Z_j^2(\xi)} \left\{ \frac{n_y Z_j^2(\xi) - G_j(\xi) V_{yj}(\xi)}{Z_j(\xi)} A_j(\xi) + G_j(\xi) W_{yj}(\xi) \right\}, \quad \text{(A3.22)}$$

$$\frac{\partial F_j(\xi)}{\partial x} = \frac{\Delta\chi_j}{T_j} \left\{ \begin{array}{l} \frac{1}{2} a_j \ln \frac{S_j(\xi)}{P_j(\xi)} - \frac{[T_j - D_j(\xi)][x(\xi) - b_j - a_j]}{S_j(\xi)} \\ - \frac{V_{xj}(\xi)}{Z_j(\xi)} A_j(\xi) - \frac{D_j(\xi)[x(\xi) - b_j]}{P_j(\xi)} - W_{xj}(\xi) \end{array} \right\}, \quad \text{(A3.23)}$$

and

$$\frac{\partial F_j(\xi)}{\partial y} = \frac{\Delta\chi_j}{T_j} \left\{ \begin{array}{l} \frac{1}{2} c_j \ln \frac{S_j(\xi)}{P_j(\xi)} - \frac{[T_j - D_j(\xi)][y(\xi) - d_j - c_j]}{S_j(\xi)} \\ - \frac{V_{yj}(\xi)}{Z_j(\xi)} A_j(\xi) - \frac{D_j(\xi)[y(\xi) - d_j]}{P_j(\xi)} - W_{yj}(\xi) \end{array} \right\}, \quad \text{(A3.24)}$$

where

$$\begin{aligned} S_j(\xi) &= T_j - 2D_j(\xi) + P_j(\xi) \\ &= [x(\xi) - b_j - a_j]^2 + [y(\xi) - d_j - c_j]^2, \end{aligned} \quad \text{(A3.25)}$$

$$\begin{aligned} V_{xj}(\xi) &= T_j[x(\xi) - b_j] - a_j D_j(\xi),\ V_{yj}(\xi) \\ &= T_j[y(\xi) - d_j] - c_j D_j(\xi), \end{aligned} \quad \text{(A3.26)}$$

$$W_{xj}(\xi) = \frac{1}{T_j}\left[\frac{a_j Z_j^2(\xi) - D_j(\xi) V_{xj}(\xi)}{P_j(\xi)} - \frac{a_j Z_j^2(\xi) + [T_j - D_j(\xi)] V_{xj}(\xi)}{S_j(\xi)}\right], \quad \text{(A3.27)}$$

and

$$W_{yj}(\xi) = \frac{1}{T_j}\left[\frac{c_j Z_j^2(\xi) - D_j(\xi) V_{yj}(\xi)}{P_j(\xi)} - \frac{c_j Z_j^2(\xi) + [T_j - D_j(\xi)] V_{yj}(\xi)}{S_j(\xi)}\right]. \quad \text{(A3.28)}$$

A3.6 Testing the Method

The algorithm discussed here to solve the exterior planar Dirichlet problem using the BEM was implemented as a program module representing a core of the FOCUS copyright software (Trubitsyn, 2010) designated for simulation of electron and ion optical systems. The experience of comprehensive testing of the software using a set of model problems can be briefly summarized with the following conclusions:

- The roundoff errors restrict the computation accuracy only in the case of electron-optical systems with electrodes possessing the rectilinear segments (Figure A3.2).
- The computation error is determined by the accuracy of approximation of a curved boundary by the rectilinear segments and does not exceed about 10^{-3} % (Figure A3.3).

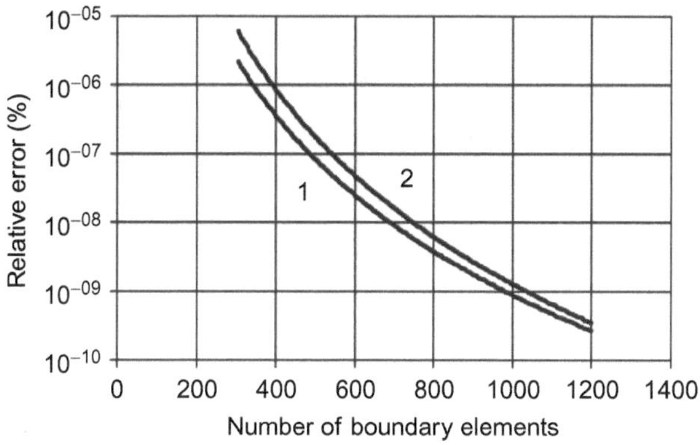

Figure A3.2 The dependence of the relative calculation error for electric potential (curve 1) and electric field (curve 2) on the number N of boundary elements in the plate capacitor on the half-distance between the plates (the ratio of the distance between the capacitor plates to their length along the Oy axis is 1:20).

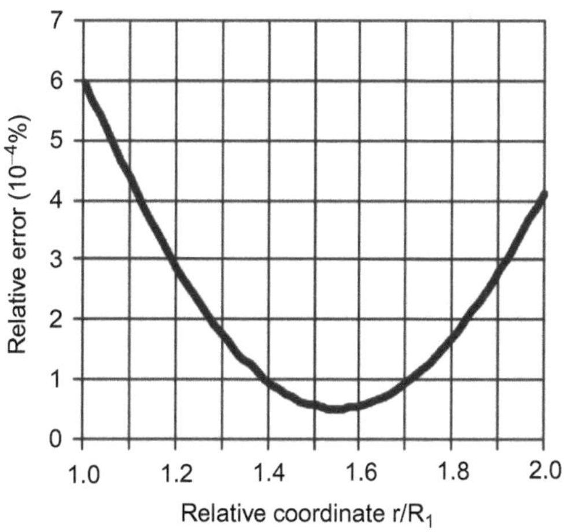

Figure A3.3 Radial dependence of the relative calculation error of the potential distribution in a cylindrical capacitor with the ratio of the radii of exterior (R_2) and interior (R_1) cylinders $R_2/R_1 = 2{:}1$. Each boundary circle is divided into 360 rectilinear elements.

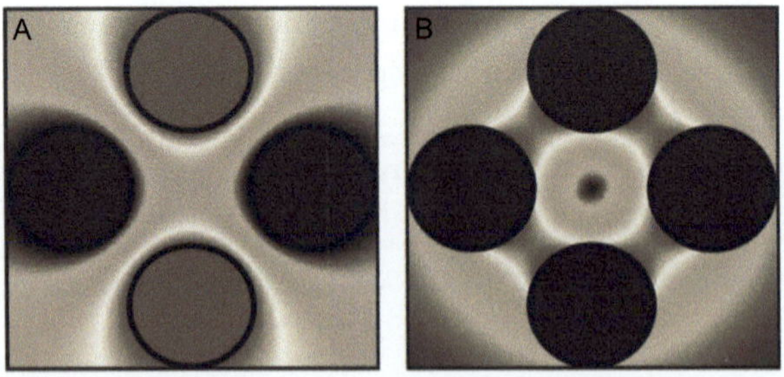

Figure A3.4 Mapping of the electrostatic potential (A) and absolute value of the potential gradient (B) in a quadrupole system. (See the color plate.)

Figure A3.4 represents an example of numerical mapping the electrostatic potential and potential gradiant in a quadrupole system, carried out with the software package "Focus".

REFERENCES

Abramovitz, M., & Stegun, I. A. (Eds.), (1964). *Handbook of mathematical functions*. Gaithersburg, MD: National Bureau of Standards.

Afanasyev, V. P., & Yavor, S. Ya. (1975a). High-luminosity energy analyzer with double focusing. *Soviet Technical Physics Letters*, *1*(9), 341–342 (translated from *Pis'ma v Zhurnal Tekhnicheskoi Fiziki*, *1*(9), 779–783).

Afanasyev, V. P., & Yavor, S. Ya. (1975b). On the possibility to increase the luminosity of energy analyzers by means of optimization of input parameters of the beam. *Zhurnal Tekhnicheskoi Fiziki*, *45*(3), 471–480 (in Russian).

Afanasyev, V. P., & Yavor, S. Ya. (1975c). Electrostatic energy analyzers. *Zhurnal Tekhnicheskoi Fiziki*, *45*(6), 1137–1170 (in Russian).

Afanasyev, V. P., & Yavor, S. Ya. (1978). *Electrostatic energy analyzers of charged particle beams*. Moscow: Nauka Publishers (in Russian).

Ashimbaeva, B. U., Chokin, K. Sh., & Saulebekov, A. O. (2003). Calculation and analysis of electron-optical characteristics of electrostatic hexapole-cylindrical field. *Proceedings of the National Academy of Sciences of the Republic of Kazakhstan, Series of Physics and Math*, *6*, 48–54.

Ashimbaeva, B. U., Chokin, K. Sh., & Saulebekov, A. O. (2005). Focusing properties of a mirror analyzer with a hexapole cylindrical field. *Journal of Electron Spectroscopy and Related Phenomena*, *143*(1), 29–32.

Ashimbaeva, B. U., Chokin, K. Sh., & Saulebekov, A. O. (2008). The focusing properties of an electrostatic mirror with a hexapole cylindrical field. *Prikladnaya Fizika (Applied Physics)*, *2*, 74–78 (in Russian).

Ashimbaeva, B. U., Chokin, K. Sh., Saulebekov, A. O., & Assylbekova, S. N. (2004). Electron optical properties of a hexapole cylindrical analyzer with face electrodes. *Eurasian Physical Technical Journal*, *1*(2), 24–29.

Ashimbaeva, B. U., Chokin, K. Sh., Saulebekov, A. O., & Kambarova, Zh. T. (2012a). The combined energy analyzer composed of electrostatic mirror fields. *Journal of Electron Spectroscopy and Related Phenomena*, *185*(11), 518–522.

Ashimbaeva, B. U., Chokin, K. Sh., Saulebekov, A. O., & Kambarova, Zh. T. (2012b). Modeling of electron-optical scheme of a hexapole-cylindrical analyzer. *Applied Physics*, *2*, 45–48 (in Russian).

Bachvalov, N. S., Zhidkov, N. P., & Kobel'kov, G. M. (1987). *Numerical methods*. Moscow: Nauka Publishers (in Russian).

Baranova, L. A., Dyakova, G. N., & Yavor, S. Ya. (1988a). Method of approximate calculation of electrostatic conical fields. *Zhurnal Tekhnicheskoi Fiziki*, *58*(1), 207–210 (in Russian).

Baranova, L. A., Dyakova, G. N., & Yavor, S. Ya (1988b). Calculation of parameters of conical systems used for analysis of charged particles in energy and angle. *Soviet Technical Physics Letters*, *33*(11), 1257–1260 (translated from *Zhurnal Tekhnicheskoi Fiziki*, *58*(11), 2069–2074).

Baranova, L. A., & Yavor, S. Ya. (1986). *Electrostatic electron lens*. Moscow: Nauka Publishers (in Russian).

Bendat, J. S., & Piersol, A. G. (1971). *Random data: Analysis and measurement procedures*. New York: John Wiley & Sons.

Brebbia, C. A., Telles, J. C. F., & Wrobel, L. C. (1984). *Boundary elements techniques (Theory and applications in engineering)*. Berlin, Heidelberg: Springer-Verlag.

Bredov, M. M. (1959). Multistage electrostatic energy analyzer with combined focusing in two planes. *Zhurnal Tekhnicheskoi Fiziki*, *29*(8), 1032–1038 (in Russian).

Cheng, A. H.-D., & Cheng, D. T. (2005). Heritage and early history of the boundary element method. *Engineering Analysis with Boundary Elements*, *29*(3), 268–302.
Courant, R., & Hilbert, D. (1951). *Methods of mathematical physics*, (Vols. 1–2). Moscow: Gostehizdat Publishers (in Russian).
Cubric, D., Kholine, N., & Konishi, I. (2011). Electron optics of spheroid charged particle energy analyzers. *Nuclear Instruments and Methods in Physics Research A*, *645*(1), 234–240.
Fishkova, T. Ya. (2012). High-dispersion charged particle energy analyzer based on two coaxial cylinders with an end-face aperture. *Zhurnal Tekhnicheskoi Fiziki*, *82*(12), 128–129 (in Russian).
Forsyte, G. E., Malcolm, M. A., & Moler, C. B. (1977). Computer methods for mathematical computations. *Prentice-Hall Series in Automatic Computation*. Englewood Cliffs, NJ: Prentice-Hall.
Franzen, W., & Taaffe, J. (1983). *Cylindrical mirror electrostatic energy analyzer free of third-order angular aberrations*. U.S. Patent 4367406 A, Washington, DC: U.S. Patent and Trademark Office.
Freykman, B. G. (1980). Separation of singularities in integral equations for a 3D electromagnetic field. *Zhurnal Tekhnicheskoi Fiziki*, *50*(2), 425–427 (in Russian).
Golikov, Yu. K., Koltsov, S. N., & Kholine, N. A. (1998). The development of high-luminosity, high-resolution electrostatic analyzer based on a cylindrical axially inhomogeneous field. *Izvestiya Akademii Nauk, Seriya Fizicheskaya*, *62*(3), 555–558 (in Russian).
Golikov, Yu. K., Koltsov, S. N., & Kholine, N. A. (1999). Implementation of a super-luminosity Auger microscope based on a new type of energy analyzer. *Izvestiya Akademii Nauk Ser Fiz*, *63*(7), 1293–1295 (in Russian).
Golikov, Yu. K., Utkin, K. U., Kholine, N. A., & Cheparuhin, V. V. (1987). *Dispersion and focusing properties of electrostatic quasiconical fields*. Preprint No. 4, Publishing House of Special Design Bureau of Analytical Instrumentation, USSR Academy of Sciences (in Russian).
Gorelik, V. A. (1986). To the theory of energy analyzers for corpuscular mass-spectroscopy. In *Abstracts of VI all-union workshop on secondary electron, photoelectron emission and solid surface spectroscopy* (pp. 190–191). Ryazan, Russia: RIC RRTI (in Russian).
Gorelik, V. (2012). Cylindrical mirror analyzer with entrance angles from 90.5° to 98.5°. *Journal of Electron Spectroscopy and Related Phenomena*, *185*(3–4), 53–54.
Gorelik, V. A., Mashinskiy, Yu. P., Pikovskaya, T. M., & Protopopov, O. D. (1985). Three-stage cylindrical mirror energy analyzer with fourth-order focusing. *Zhurnal Tekhnicheskoi Fiziki*, *55*(2), 412–414 (in Russian).
Gorelik, V. A., Mashinskiy, Yu. P., Protopopov, O. D., & Shapiro, A. A. (1976). Choice of the design for a cylindrical energy analyzer and numerical analysis of instrumental distortions of Auger spectra. *Izvestiya Academy of Sciences USSR. Physical Series*, *40*(12), 2633–2636 (in Russian).
Gorelik, V. A., Protopopov, O. D., & Trubitsyn, A. A. (1988). Search for high-order focusing in real electron-optical systems. *Zhurnal Tekhnicheskoi Fiziki*, *58*(8), 1531–1534 (in Russian).
Greenberg, M. D. (1971). *Application of Green's functions in science and engineering*. Englewood Cliffs, NJ: Prentice Hall.
Greenfield, D. E., & Monastyrskiy, M. A. (2009). Selected problems of computational charged particle optics. In P. W. Hawkes (Ed.), *Advances in imaging and electron physics: Vol. 155* (pp. 1–32). London: Academic Press. Chapter 1.
Gurov, V. S. (1976). Energy analyzer of charged particles on the axially symmetric quadrupole lens. *Electronics: Interuniversity collection of scientific papers: Vol. 3*. (pp. 77–79). Ryazan, Russia: RIC RRTI (in Russian).
Gurov, V. S. (1977). Energy analyzer of charged particles. *Electronics: Interuniversity collection of scientific papers: Vol. 4*. (pp. 133–135). Ryazan, Russia: RIC RRTI (in Russian).

Gurov, V. S. (1982). Experimental research of the ion-optic system of the laser-plasma source of ions. *Vacuum and plasma electronics: Interuniversity collection of scientific papers* (pp. 134–137). Ryazan, Russia: RIC RRTI (in Russian).

Gurov, V. S. (1997). Energy analyzers of charged particles on the axially symmetric hyperboloid systems. In *Abstracts of International scientific and technical conference "Scientific basis of high technologies, Novosibirsk* (pp. 67–70) (in Russian).

Gurov, V. S., & Kuksa, P. I. (2013). Hyperboloid axially symmetric energy analyzer with the conic end electrode. *Vestnik Ryazanskogo Radiotehnicheskogo Universiteta (Journal of the Ryazan State Radio Engineering University), 46*(4), 73–76 (in Russian).

Gurov, V. S., & Kuksa, P. I. (2014a). Hyperboloid axially symmetric energy analyzer on the simplified electrode system. *Vestnik Ryazanskogo Radiotehnicheskogo Universiteta (Journal of the Ryazan State Radio Engineering University), 47*(3), 110–113 (in Russian).

Gurov, V. S., & Kuksa, P. I. (2014b). Hyperboloid axially symmetric energy analyzer on two annular electrodes. *Uspehi Sovremennoj Radioelectroniki (Successes of Contemporary Radio Electronics), 6*, 81–85 (in Russian).

Gurov, V. S., & Sheretov, E. P. (1984). On the possibility of designing a charged particle energy analyzer on the basis of a hyperboloidal axisymmetric lens. *Zhurnal Tekhnicheskoi Fiziki, 54*(12), 2383–2386 (in Russian).

Gurov, V. S., Sheretov, E. P., & Sakhanova, V. V. (1985). *Energy analyzer of charged particles*. USSR Patent 1246174, Washington, DC: U.S. Patent and Trademark Office.

Gurov, V. S., Trubitsyn, A. A., Grachev, E. Yu, & Kuksa, P. I. (2015). About the focusing of charged particles by electrostatic axially symmetric hyperbolic field. *Radiotekhnika (Radio Engineering), 11*. (in print) (in Russian).

Gurov, V. S., Trubitsyn, A. A., & Kuksa, P. I. (2015). Influence of geometry of electrode system on efficiency of hyperboloid energy analyzers. *Radiotekhnika (Radio Engineering), 5*, 60–65 (in Russian).

Huges, A. L., & Rojansky, V. (1929). On the analysis of electronic velocities by electrostatic means. *Physical Review, 34*(4), 284–290.

Jacka, M., Kale, A., & Traitler, N. (2003). Hyperbolic field electron energy analyzer with second-order focusing. *Review of Scientific Instruments, 74*(10), 4298–4300.

Kanwal, R. P. (1996). *Linear integral equations: Theory and technique* (2nd ed.). Boston: Birkhäuser.

Khursheed, A., Hoang, H. Q., & Srinivasan, A. (2012). A wide-range parallel radial mirror analyzer for scanning electron/ion microscopes. *Journal of Electron Spectroscopy and Related Phenomena, 184*(11–12), 525–532.

Kover, A., Varga, D., Szabo, Gy, Berenyi, D., Kadar, I., Ricz, S., et al. (1983). A study of the angular distribution of the electrons in the peak near ve = vi in the electron spectra from He^+-Ar and H_2^+-Ar collisions. *Journal of Physics B, 16*(10), 1017–1021.

Kozlov, I. G. (1971). *Methods of the energy-dispersive analysis of electron flows*. Moscow: Atomizdat Publishers (in Russian).

Kozlov, I. G. (1978). *Modern problems of electron spectroscopy*. Moscow: Atomizdat Publishers (in Russian).

Leckey, R. C. G. (1987). Recent developments in electron energy analyzers. *Journal of Electron Spectroscropy and Related Phenomena, 43*(3), 183–214.

Manura, D. J., & Dahl, D. A. (2006). *SIMION 8.0 user manual*. Idaho Falls, ID: Scientific Instrument Services Inc., Idaho National Laboratory.

Menshikov, K. A. (1982). An electrostatic analyzer of charged particles with three coaxial cylindrical electrodes. *Zhurnal Tekhnicheskoi Fiziki, 52*(11), 2245–2252 (in Russian).

Mikheeva, E. V., & Protopopov, O. D. (1982). Methods of electron and ion spectroscopy. *Electronnaya Promyshlennost (Electronic Industry), 10–11*, 10–15 (in Russian).

Mollenstedt, G. (1949). Die electrostatische Linse als hochauflosender Geschwindigkeitanalysator. *Optik, 5*, 499.

Morton, K. W., & Mayers, D. F. (2005). *Numerical solution of partial differential equations: An introduction*. Cambridge, UK: Cambridge University Press.

Ovsyannikova, L. P., & Yavor, S. Ya. (1978). Axisymmetric electrostatic energy analyzers of charged particles. *Zhurnal Tekhnicheskoi Fiziki*, *48*(6), 1306–1308 (in Russian).

Purcell, E. M. (1938). The focusing of charged particles by a spherical condenser. *Physics Review*, *54*(10), 813–826.

Samarskiy, A. A. (1983). *The theory of difference schemes*. Moscow: Nauka Publishers (in Russian).

Sar-El, H. Z. (1966). More on the spherical condenser as an analyzer, I. Nonrelativistic part. *Nuclear Instruments and Methods*, *42*(1), 71–76.

Schmidt, C. A. (1970). Quadrupole energy filter. *Review of Scientific Instruments*, *41*(1), 117–121.

Seah, M. P., & Briggs, D. (1992). *Practical surface analysis by auger and X-ray photoelectron spectroscopy* (2nd ed.). Chichester, UK: Wiley & Sons.

Shapovalov, A. S. (1968). On focusing and dispersive properties of the elliptic electron mirror. *Zhurnal Tekhnicheskoi Fiziki*, *38*(12), 2087–2092 (in Russian).

Sheretov, E. P., & Gurov, V. S. (1980). *Sensor of the hyperbolic mass-spectrometer*. USSR Patent 951477.

Sheretov, E. P., & Gurov, V. S. (1982). *Laser-plasma source of ions*. USSR Patent 1045778.

Sheretov, E. P., & Gurov, V. S. (1985). On selection of the parameter determining the effectiveness of the hyperboloidal axisymmetric energy analyzers. *Zhurnal Tekhnicheskoi Fiziki*, *55*(8), 1632–1635 (in Russian).

Sheretov, E. P., Gurov, V. S., Dubkov, M. V., & Korneeva, O. V. (1999). The monopole mass filter with hyperbolic V-shaped electrode. *Rapid Communications in Mass Spectrometry*, *13*, 1699–1702.

Sheretov, E. P., Gurov, V. S., Safonov, M. P., & Philippov, I. V. (1999). Hyperboloid mass spectrometers for space exploration. *International Journal of Mass Spectrometry*, *189*, 9–17.

Sheretov, E. P., Safonof, M. P., & Gurov, V. S. (1987). Hyperboloidal axisymmetric energy analyzer with particle return. *Zhurnal Tekhnicheskoi Fiziki*, *57*(6), 1185–1188 (in Russian).

Sheretov, E. P., Zenkin, V. A., Safonov, M. P., Gurov, V. S., & Sakhanova, V. V. (1985). *Ion micro analyzer*. USSR Patent 1191981.

Siegbahn, K., Kholine, N., & Golikov, Yu. (1997). A high-resolution and large-transmission electron spectrometer. *Nuclear Instruments and Methods in Physics Research*, *A384*, 563–574.

Sternberg, W. J., & Smith, T. L. (1944). *The theory of potential and spherical harmonics*. Toronto: University of Toronto Press.

Trubitsyn, A. A. (1995a). Calculation of the singular intergrals arising in the boundary-element method for the Dirichlet problem. *Computational Mathematics and Mathematical Physics*, *35*(4), 421–428.

Trubitsyn, A. A. (1995b). Cylindrical mirror analyzer with high energy resolution. *Journal of Electron Spectroscopy and Related Phenomena*, *73*, 305–310.

Trubitsyn, A. A. (2000). A correlation method of search for higher-order angular focusing. *Technical Physics*, *46*(5), 126–127.

Trubitsyn, A. A. (2008). The software "FOCUS" modelling axi-symmetric electron-optical systems: Algorithms and characteristics. *Applied Physics*, *2*, 56–61 (in Russian).

Trubitsyn, A. A. (2010). The software "FOCUS" to simulate axi-symmetrical and planar electron (ion) optical systems. In *Abstracts of the eighth international conference on charged particle optics (CPO-8). Software demo session. Singapore*, pp. 208–209.

Varga, D., Tokesi, K., & Rajta, I. (1995). Design of an electrostatic electron spectrometer for simultaneous energy and angular distribution measurements. *Journal of Electron Spectroscopy and Related Phenomena*, *76*, 433–436.

von Zahn, U. (1963). Monopole spectrometer, a new electric field mass spectrometer. *Review of Scientific Instruments*, *34*, 1–4.

Woodruff, D. P., & Delchar, T. A. (1986). *Modern techniques of surface science*. Cambridge, UK: Cambridge University Press.

Wrobel, L. C., & Aliabadi, M. H. (2002). *The boundary element method*. New York: John Wiley & Sons.

Zashkvara, V. V. (1995). On circular multipole calculations. *Nuclear Instruments and Methods in Physics Research Section A*, *A354*, 171–174.

Zashkvara, V. V., Ashimbaeva, B. U., & Chokin, K. Sh. (2002). Calculation of trajectories in a multipole cylindrical field. *Journal of Electron Spectroscopy and Related Phenomena*, *122*, 195–202.

Zashkvara, V. V., Ashimbaeva, B. U., Chokin, K. Sh., & Rysavy, M. (1992). Effect of the fringe field of one-dimensional aperture grating on the transmission and focusing of the cylindrical mirror analyzer. *Journal of Electron Spectroscopy and Related Phenomena*, *58*, 271–278.

Zashkvara, V. V., Ashimbaeva, B. U., Saulebekov, A. O., & Tajibaeva, S. D. (1994). Theoretical justification of a spectrograph for photoelectron spectroscopy with angular resolution. *Proceedings of the National Academy of Sciences of the Republic of Kazakhstan, Series of Physics and Math*, *2*, 75–81.

Zashkvara, V. V., Ilyin, A. M., & Kruchkov, V. F. (1977). Two cases of focusing of axisymmetric beam of charged particles in electrostatic hyperboloidal field. *Zhurnal Tekhnicheskoi Fiziki*, *47*(7), 1572–1574 (in Russian).

Zashkvara, V. V., Korsunskii, O. S., & Kosmachev, O. S. (1966). Focusing properties of an electrostatic mirror with the cylindrical field. *Soviet Technical Physics Letters*, *11*, 96–99 (translated from *Zhurnal Tekhnicheskoi Fiziki*, *36*(1), 132–138).

Zashkvara, V. V., & Red'kin, V. S. (1969). To the question on focusing a beam of charged particles by electrostatic mirror in cylindrical field. *Zhurnal Tekhnicheskoi Fiziki*, *39*(8), 1452–1456 (in Russian).

Zashkvara, V. V., & Tyndyk, N. N. (1996). Potential fields based on circular multipole series. *Nuclear Instruments and Methods in Physics Research Section A*, *A370*, 452–460.

Zashkvara, V. V., & Tyndyk, N. N. (1999). The method for the calculation of multipole-cylindrical fields. *Nuclear Instruments and Methods in Physics Research Section A*, *A370*, 223–231.

Zashkvara, V. V., Yurchak, L. S., & Bylinkin, A. F. (1988). Electron-optical properties of electrostatic spherical mirror and systems on its basis (I). *Zhurnal Tekhnicheskoi Fiziki*, *58*(10), 2010–2020 (in Russian).

Zienkiewicz, O. C., Taylor, R. L., & Zhu, J. Z. (2005). *The finite element method: Its basis and fundamentals*. Oxford: Butterworth-Heinemann.

Zigban, K., et al. (1971). *Electron spectroscopy*. Moscow: Mir Publishers (in Russian).

CONTENTS OF VOLUMES 151-191

Volume 151[1]

Claas Bontus and Thomas Köhler, Reconstruction algorithms for computed tomography

Laurent Busin, Nicolas Vandenbroucke and Ludovic Macaire, Color spaces and image segmentation

Glenn R. Easley and Flavia Colonna, Generalized discrete Radon transforms and applications to image processing

Tomáš Radlička, Lie agebraic methods in charged particle optics

Valerie Randle, Recent developments in electron backscatter diffraction

Volume 152

Nina S.T. Hirata, Stack filters: from definition to design algorithms

Sameen Ahmed Khan, The Foldy–Wouthuysen transformation technique in optics

Saverio Morfu, Patrick Marquié, Brice Nofiélé and Dominique Ginhac, Nonlinear systems for image processing

Tohru Nitta, Complex-valued neural network and complex-valued backpropagation learning algorithm

Jérôme Bobin, Jean-Luc Starck, Y. Moudden and M.J. Fadili, Blind source separation: the sparsity revoloution

Ray L. Withers, "Disorder": structured diffuse scattering and local crystal chemistry

Volume 153

Aberration-corrected Electron Microscopy

Harald Rose, History of direct aberration correction

Maximilian Haider, Heiko Müller and Stephan Uhlemann, Present and future hexapole aberration correctors for high-resolution electron microscopy

Ondrej L. Krivanek, Niklas Dellby, Robert J. Kyse, Matthew F. Murfitt, Christopher S. Own and Zoltan S. Szilagyi, Advances in aberration-corrected scanning transmission electron microscopy and electron energy-loss spectroscopy

Philip E. Batson, First results using the Nion third-order scanning transmission electron microscope corrector

Andrew L. Bleloch, Scanning transmission electron microscopy and electron energy loss spectroscopy: mapping materials atom by atom

Florent Houdellier, Martin Hÿtch, Florian Hüe and Etienne Snoeck, Aberration correction with the SACTEM-Toulouse: from imaging to diffraction

Bernd Kabius and Harald Rose, Novel aberration correction concepts

Angus I. Kirkland, Peter D Nellist, Lan-yun Chang and Sarah J. Haigh, Aberration-corrected imaging in conventional transmission electron microscopy and scanning transmission electron microscopy

Stephen J Pennycook, M.F. Chisholm, A.R. Lupini, M. Varela, K. van Benthem, A.Y. Borisevich, M.P. Oxley, W. Luo and S.T. Pantelides, Materials applications of aberration-corrected scanning transmission electron microscopy

Nobuo Tanaka, Spherical aberration-corrected transmission electron microscopy for nanomaterials

Knut Urban, Lothar Houben, Chun-lin Jia, Markus Lentzen, Shao-bo Mi, Andreas Thust and Karsten Tillmann, Atomic-resolution aberration-corrected transmission electron microscopy

Yimei Zhu and Joe Wall, Aberration-corrected electron microscopes at Brookhaven National Laboratory

[1] Lists of the contents of volumes 100–149 are to be found in volume 150; the entire series can be searched on ScienceDirect.com

Volume 154

Henning F. Harmuth and Beate Meffert, Dirac's Difference Equation and the Physics of Finite Differences

Volume 155

Dmitry Greenfield and Mikhail Monastyrskiy, Selected Problems of Computational Charged Particle Optics

Volume 156

Vasileios Argyriou and Maria Petrou, Photometric stereo: an overview
Fred Brackx, Nele de Schepper and Frank Sommen, The Fourier transform in Clifford analysis
Niels de Jonge, Carbon nanotube electron sources for electron microscopes
Erasmo Recami and Michel Zamboni-Rached, Localized waves: a review

Volume 157

Mikhail I. Yavor, Optics of charged particle analyzers

Volume 158

Péter Dombi, Surface plasmon-enhanced photoemission and electron acceleration with ultrashort laser pulses
Brian J. Ford, Did physics matter to the pioneers of microscopy?
Jérôme Gilles, Image decomposition: theory, numerical schemes, and performance evaluation
Stina Svensson, The reverse fuzzy distance transform and its use when studying the shape of macromolecules from cryo-electron tomographic data
Marc van Droogenbroeck, Anchors of morphological operators and algebraic openings
Dong Yang, Shiva Kumar and Hao Wang, Temporal filtering technique using time lenses for optical transmission systems

Volume 159

Cold Field Emission and the Scanning Transmission Electron Microscope
Albert Victor Crewe, The work of Albert Victor Crewe on the scanning transmission electron microscope and related topics
Lyn W. Swanson and Gregory A. Schwind, A review of the cold-field electron cathode
Joseph S. Wall, Martha N. Simon and James F. Hainfeld, History of the STEM at Brookhaven National Laboratory
Hiromi Inada, Hiroshi Kakibayashi, Shigeto Isakozawa, Takahito Hashimoto, Toshie Yaguchi and Kuniyasu Nakamura, Hitachi's development of cold-field emission scanning transmission electron microscopes
Peter W. Hawkes, Two commercial STEMs: the Siemens ST100F and the AEI STEM-1
Ian R.M. Wardell and Peter E. Bovey, A history of Vacuum Generators' 100-kV STEM
H. Sebastian von Harrach, Development of the 300-kV Vacuum Generators STEM (1985–1996)
Bernard Jouffrey, On the high-voltage STEM project in Toulouse (MEBATH)
Andreas Engel, Scanning transmission electron microscopy: biological applications
Kenneth C.A. Smith, STEM at Cambridge University: reminiscences and reflections from the 1950s and 1960s

Volume 160

Zofia Baranczuk, Joachim Giesen, Klaus Simon and Peter Zolliker, Gamut mapping
Adrian N. Evans, Color area morphology scalespaces
Ye Pu, Chia-lung Hsieh, Rachel Grange and Demetri Psaltis, Harmonic holography
Gerhard X. Ritter and Gonzalo Urcid, Lattice algebra approach to endmember determination in hyperspectral imagery
Reinhold Rüdenberg, Origin and background of the invention of the electron microscope
H. Gunther Rudenberg and Paul G. Rudenberg, Origin and background of the invention of the electron microscope: commentary and expanded notes on Memoir of Reinhold Rüdenberg

Volume 161

Marian Mankos, Vassil Spasov and Eric Munro, Principles of dual-beam low-energy electron microscopy
Jorge D. Mendiola-Santibañez, Iván R. Terol-Villalobos and Israel M. Santillán-Méndez, Determination of adequate parameters for connected morphological contrast mappings through morphological contrast measures
Ignacio Moreno and Carlos Ferreira, Fractional Fourier transforms and geometrical optics
Vladan Velisavlevic, Martin Vetterli, Baltasar Berufell-Lozano and Pier Luigi Dragotti, Sparse image representation by directionlets
Michael H.F. Wilkinson and Georgios K. Ouzounis, Advances in connectivity and connected attribute filters

Volume 162

Kiyotaka Asakura, Hironobu Niimi and Makoto Kato, Energy-filtered x-ray photoemission electron microscopy (EXPEEM)
Eireann C. Cosgriff, Peter D. Nellist, Adrian J. d'Alfonso, Scott D. Findlay, Gavin Behan, Peng Wang, Leslie J. Allen and Angus I. Kirkland, Image contrast in aberration-corrected scanning confocal electron microscopy
Christopher J. Edgcombe, New dimensions for field emission: effects of structure in the emitting surface
Archontis Giannakidis and Maria Petrou, Conductivity imaging and generalised Radon transform: a review
Olivier Losson, Ludovic Macaire and Yanqin Yang, Comparison of color demosaicing methods

Volume 163

Wolfgang S. Bacsa, Optical interference near surfaces and its application in sub-wavelength microscopy
Ruy H.A. Farias and Erasmo Recami, Introduction of a quantum of time ("chronon"), and its consequences for the electron in quantum and classical physics
Andrew Neice, Methods and limitations of subwavelength imaging
A. Sever Škapin and P. Ropret, Identification of historical pigments in wall layers by combination of optical and scanning electron microscopy coupled with energy-dispersive spectroscopy
Markus E. Testorf and Michael A. Fiddy, Superresolution imaging–revisited

Volume 164

Amos Bardea and Ron Naaman, Magneto-lithography: from the bottom-up route to high throughput
Román Castañeda, The optics of spatial coherence wavelets
Junchang Li, Yanmei Wu and Yan Li, Common diffraction integral calculation based on a fast Fourier transform algorithm
Marcel Teschke and Stefan Sinzinger, A generalized approach to describe the interference contrast and phase contrast method
Dokkyu Yi and Booyong Choi, Nonlinear partial differential equations for noise problems
Henning F. Harmuth, Harmuth corrigenda

Volume 165

Natalie Baddour, Two-dimensional Fourier transforms in polar coordinates
Neil V. Budko, Superluminal, subluminal, and negative velocities in free-space electromagnetic propagation
Rowan Leary and Rik Brydson, Chromatic aberration correction: the next step in electron microscopy
Michele Marrocco, Methods for vectorial analysis and imaging in high-resolution laser microscopy
Tomoya Sakai, Masaki Narita, Takuto Komazaki, Haruhiko Nishiguchi and Atsushi Imiya, Image hierarchy in Gaussian scale space
Yusuf Ziya Umul, The theory of the boundary diffraction wave
Emil Wolf, History and solution of the phase problem in the theory of structure determination of crystals from x-ray diffraction measurements

Volume 166

Valeriy Syrovoy, Theory of Intense Beams of Charged Particles

Volume 167

Emmanuel de Chambost, A history of Cameca (1954–2009)
Johan Debayle and Jean-Charles Pinoli, Theory and applications of general adaptive neighborhood image processing
Mohamed ben Haj Rhouma, Mohamed Ali Khabou and Lotfi Hermi, Shape recognition based on eigenvalues of the Laplacian
Nicolas Loménie and Georges Stamon, Point set analysis
Leonid P. Yaroslavsky, Image recovery from sparse samples, discrete sampling theorem, and sharply bounded band-limited discrete signals

Volume 168

Luca Geretti and Antonio Abramo, The synthesis of a stochastic artificial neural network application using a genetic algorithm approach
Michel Jourlin, Josselin Breugnot, Frédéric Itthirad, Mohammed Bouabdellah and Brigitte Closs, Logarithmic image processing for color images
Rainer A. Leitgeb, Current technologies for high-speed and functional imaging with optical coherence tomography
Sergej A. Nepijko and Gerd Schönhense, Analysis of optical systems, contrast depth, and measurement of electric and magnetic field distribution on the object's surface in mirror electron microscopy
Chad M. Parish, Multivariate statistics applications in scanning transmission electron microscopy
Hidetaka Sawada, Fumio Hosokawa, Takeo Sasaki, Toshikatsu Kaneyama, Yukihito Kondo and Kazutomo Suenaga, Aberration correctors developed under the Triple C project
Tobias Schulz, Martin Albrecht and Klaus Irmscher, Spatially resolved thermoluminescence in a scanning electron microscope

Volume 169

Erchan Aptoula and Sébastien Lefèvre, Morphological texture description of grayscale and color images
Vera Guarrera and Herwig Ott, Electron microscopy of ultracold gases
Konstantinos Konstantinidis, Ioannis Andreadis and Georgios Ch. Sirakoulis, Application of artificial intelligence to content-based image retrieval
Xingwei Yang, Daniel B. Szyld and Longin Jan Latecki, Diffusion on a tensor product graph for semi-supervised learning and interactive image segmentation
S.A. Nepijko and G. Schönhense, Electron holography for electric and magnetic field measurement and its application for nanophysics

Volume 170

Alex S. Eggeman and Paul A. Midgley, Precession electron diffraction
Ray Hill, John A. Notte and Larry Scipione, Scanning helium ion microscopy
Hone-Ene Hwang and Pin Han, Signal reconstruction algorithm based on a single intensity in the Fresnel domain
Kazuhisa Sato, Toyohiko J. Konno and Yoshihiko Hirotsu, Electron microscpy studies on magnetic $L1_0$ FePd nanoparticles
D.A. Zanin, H. Cabrera, L. de Pietro, M. Pikulski, M. Goldmann, U. Ramsperger, D. Pescia and John P. Xanthakis, Fundamental aspects of near-field emission scanning electron microcopy

Volume 171

Gregor Esser, Wolfgang Becken, Werner Müller, Peter Baumbach, Josep Arasa and Dietmar Uttenweiler, Derivation of the reflection equations for higher order aberrations of local wavefronts by oblique incidence
Lila Iznita Izhar and Maria Petrou, Thermal imaging in medicine
Jean-Michel Tualle, Derivation of the radiative transfer equation in a medium with a spatially varying refractive index: a review
Kamlesh Shrivas and Mitsutoshi Setou, Imaging mass spectrometry. Sample preparation, instrumentation and applications
Robert T. Thompson and Steven A. Cummer, Transformation optics

Tobias Klein, Egbert Buhr and Carl Georg Frase, TSEM – a review of scanning electron microscopy in transmission mode and its applications
Michel Jourlin, Maxime Carré, Josselin Breugnot and Mohamed Bouabdellah, Logarithmic image procesing: additive contrast, multiplicative contrast and associated metrics
Partha Pratim Mondal and Alberto Diaspro, Point spread function engineering for super-resolution single-photon and multiphoton fluorescence microscopy
Paul Murray and Stephen Marshall, A review of recent advances in the hit-or-miss transform
Stephen J. Sangwine, Perspectives on color image procesing by linear vector methods using projective geometric transformations

Volume 172

Jay Theodore Cremer, Neutron and x-ray microscopy, Part 1

Volume 173

Jay Theodore Cremer, Neutron and x-ray microscopy, Part 2

Volume 174

Silicon-based Millimeter-wave Technology. Measurement, Modeling and Applications
M. Jamal Deen and Ognian Marinov, Measurement techniques and issues
Guennadi A. Kouzaev, M. Jamal Deen and Natalia K. Nikolova, Transmission lines and passive components
Mohamed H. Bakr and Mohamed H. Negm, Modeling and design of high-frequency structures using artificial neural networks and space mapping
Oana Moldovan, Antonio Lázaro, François Danneville, Rodrigo Picos, Bogdan Nae, Benjamin Iniguez and M. Jamal Deen, Nanoscale FETs
M. Daneshmand and R.R. Mansour, RF MEMS switches and switch matrices
Natalia K. Nikolova, Maryam Ravan and Reza K. Amineh, Substrate-integrated antennas on silicon

Volume 175

Jay Theodore Cremer, Small angle scatter with correlation, scatter and intermediate functions
Jay Theodore Cremer, Nuclear scatter of neutron spin states
Christian Dwyer, Atomic-resolution core-level spectroscopy in the scanning transmission electron microscope

Volume 176

Katsushige Tsuno, Damaschin Ioanoviciu, Early History of Wien Filters
Damaschin Ioanoviciu, Katsushige Tsuno, Aberration Theory of the Wien Filter
Katsushige Tsuno, Damaschin Ioanoviciu, Wien Filter Instrumentation
Katsushige Tsuno, Damaschin Ioanoviciu, Simulation of Multipole Wien Filters
Damaschin Ioanoviciu, Katsushige Tsuno, Wien Filter Applications to Ions
Katsushige Tsuno, Damaschin Ioanoviciu, Application of Wien Filters to Electrons

Volume 177

Michel Jourlin , Josselin Breugnot, Bassam Abdallah, Joris Corvo, Enguerrand Couka , Maxime Carré, Image Segmentation in the Field of the Logarithmic Image Processing Model: Special Focus on the Hierarchical Ascendant Classification Techniques
Petros Maragos, Representations for Morphological Image Operators and Analogies with Linear Operators
Kenneth C. A. Smith Electron Microscopy at Cambridge University with Charles Oatley and Ellis Cosslett: Some Reminiscences and Recollections
Miguel José-Yacamán, Arturo Ponce, Sergio Mejía-Rosales, Francis Leonard Deepak, Advanced Methods of Electron Microscopy in Catalysis Research

Volume 178

Tony Lindeberg, Generalized Axiomatic Scale-Space Theory
Agnieszka Lisowska, Smoothlet Transform: Theory and Applications

Evgeniy M. Yakushev, Theory and Computation of Electron Mirrors: The Central Particle Method

Volume 179

Claude Daviau, Invariant Quantum Wave Equations and Double Space-Time
Niels de Jonge, In-Situ and Correlative Electron Microscopy
Vladimir P. Oleshko, James M. Howe, Electron Tweezers as a Tool for High-Precision Manipulation of Nanoobjects
Pilar Sobrevilla, Eduard Montseny, Aina Barcelo, Robustness Analysis of the Reduced Fuzzy Texture Spectrum and its Performance on Noisy Images
Arturo Tejada, Wouter Van den Broek, Arnold J. den Dekker, Measure-by-Wire (MBW): An Automatic Control Framework for High-Throughput Transmission Electron Microscopy

Volume 180

Mikhail Ya. Schelev, Mikhail A. Monastyrskiy, Nikolai S. Vorobiev, Sergei V. Garnov and Dmitriy E. Greenfield, Aspects of Streak Image Tube Photography

Volume 181

Ying Bai, Xiao Han, Jerry L. Prince, Octree Grid Topology-Preserving Geometric Deformable Model (OTGDM)
Maïtine Bergounioux, Second-order Variational Models for Image Texture Analysis
Victoria Klang, Nadejda B. Matsko, Electron Microscopy of Pharmaceutical Systems

Volume 182

Hans R. Gelderblom, Detlev H. Krüger, Helmut Ruska (1908–1973): His Role in the Evolution of Electron Microscopy in the Life Sciences, and Especially Virology
Hans R. Gelderblom, Detlev H. Krüger, Peter W. Hawkes Publications from the Düsseldorf University Institute for Biophysics and Electron Microscopy (Institut für Biophysikund Elektronenmikroskopie der Universität Düsseldorf) 1958-1973
Nebojsa Neškovič, P. Beličev, I. Telečki, S. Petrovič, Rainbow Lenses
Ben Adcock, Anders Hansen, Bogdan Roman, Gerd Teschke, Generalized Sampling: Stable Reconstructions, Inverse Problems and Compressed Sensing over the Continuum

Volume 183

M.M. El-Gomati, C.G.H. Walker, Toward Quantitative Scanning Electron Microscopy
Laurent Navarro, Guy Courbebaisse, Michel Jourlin, Logarithmic Wavelets
F. Lanusse, J.-L. Starck , A. Woiselle, M. J. Fadili, 3-D Sparse Representations

Volume 184

Anatoli A. Ischenko, Sergei A. Aseyev, Time-Resolved Electron Diffraction: for Chemistry, Biology and Materials Science

Volume 185

Pawel Berczynski, Slawomir Marczynski, Gaussian Beam Propagation in
Inhomogeneous Nonlinear Media. Description in Ordinary Differential Equations by Complex Geometrical Optics
David Agard, Yifan Cheng, Robert M. Glaeser, Sriram Subramaniam, Single-Particle Cryo-Electron Microscopy (Cryo-EM): Progress, Challenges, and Perspectives for Further Improvement
Martin Welk, Michael Breuß, Morphological Amoebas and Partial Differential Equations

Volume 186

Niels de Jonge, Marina Pfaff, Diana B. Peckys
Practical Aspects of Transmission
Electron Microscopy in Liquid
Jian-Jiun Ding, Soo-Chang Pei
Linear Canonical Transform
Andrey I. Denisyuk, Alexey V. Krasavin, Filipp E. Komissarenko, Ivan S. Mukhin

Mechanical, Electrostatic, and Electromagnetic Manipulation of Microobjects and Nanoobjects in Electron Microscopes

Volume 187

Ahmed Elgammal, Homeomorphic Manifold Analysis (HMA): Untangling Complex Manifolds

Teruo Kohashi, Spin-Polarized Scanning Electron Microscopy

Volume 188

Allen M. Carroll, Pattern Generators for Reflective Electron-Beam Lithography (REBL)

Frank Gunzer, Jürgen Grotemeyer, Recent Developments in Time-of-Flight Mass Spectrometry

Margit Pap, A Special Voice Transform, Analytic Wavelets, and Zernike Functions

Colin J.R. Sheppard, Shan S. Kou, Jiao Lin, The Hankel Transform in *n*-dimensions and Its Applications in Optical Propagation and Imaging

Volume 189

Georges Lochak, Theory of the Leptonic Monopole, Part 1

Harald Stumpf, Symmetry Breaking by Electric Discharges in Water and Formation of Lochak's Light Magnetic Monopoles in an Extended Standard Model, Part 2

Volume 190

Niels de Jongek, CISCEM 2014: Proceedings of the Second Conference on In situ and Correlative Electron Microscopy, Saarbrücken, Germany, October 14–15, 2014

A. R. Faruqi, Richard Henderson, and Greg McMullan, Progress and Development of Direct Detectors for Electron Cryomicroscopy

Peter W. Hawkes, Electron Optics and Electron Microscopy Conference Proceedings and Abstracts: A Supplement

Grzegorz Wielgoszewski and Teodor Gotszalk, Scanning Thermal Microscopy (SThM): How to Map Temperature and Thermal Properties at the Nanoscale

Volume 191

Martin Berz, Philip M. Duxbury, Kyoko Makino, and Chong-Yu Ruan, Femtosecond Electron Imaging and Spectroscopy

Ronald E. Burge, Imaging with Electrons, X-rays, and Microwaves: Some Scattered Thoughts

INDEX

Note: Page numbers followed by "*f*" indicate figures and "*t*" indicate tables.

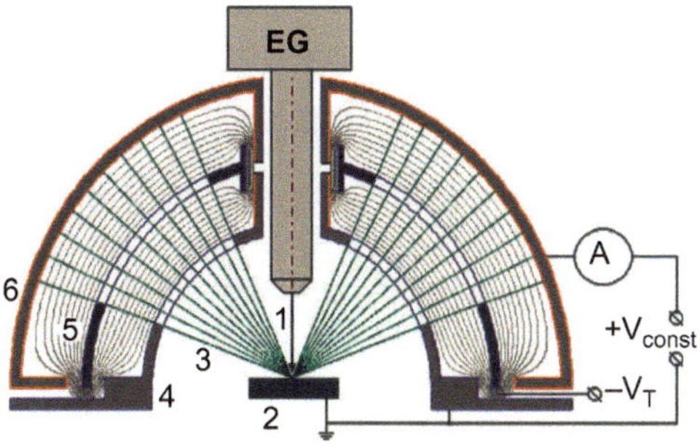

Chapter 1, Figure 1 Quasi-spherical capacitor scheme: 1—exciting radiation, 2—sample, 3—secondary particles, 4—first grid, 5—retarding grid, 6—collector, EG—electron gun.

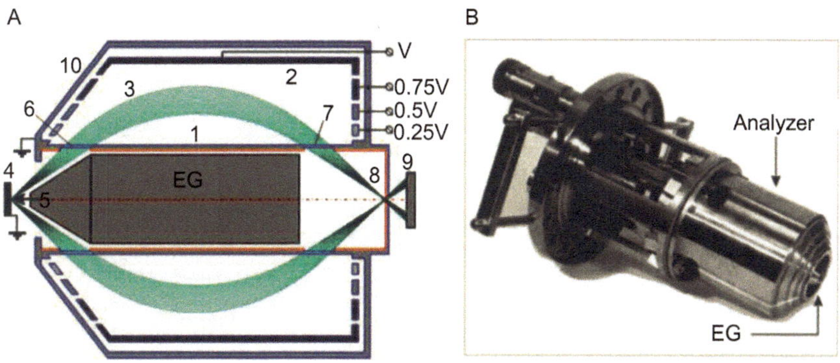

Chapter 1, Figure 3 Electron-optical scheme (A) and external view (NITI, Ryazan) (B) of the cylindrical mirror energy analyzer: 1—inner cylinder, 2—outer cylinder, 3—secondary electrons, 4—sample, 5—primary electrons, 6—input window 7—output window, 8—diaphragm 9—collector (SEM), 10—magnetic shield, EG—electron gun.

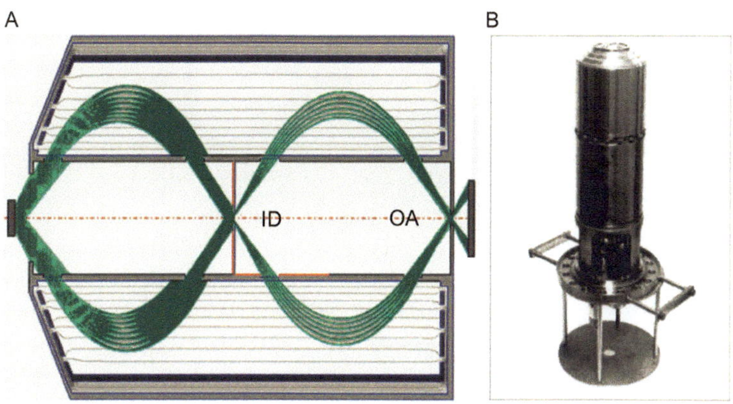

Chapter 1, Figure 4 Electron-optical scheme (A) and external view (NITI, Ryazan) (B) of the two-stage cylindrical mirror energy analyzer: ID—intermediate diaphragm, OA—output aperture.

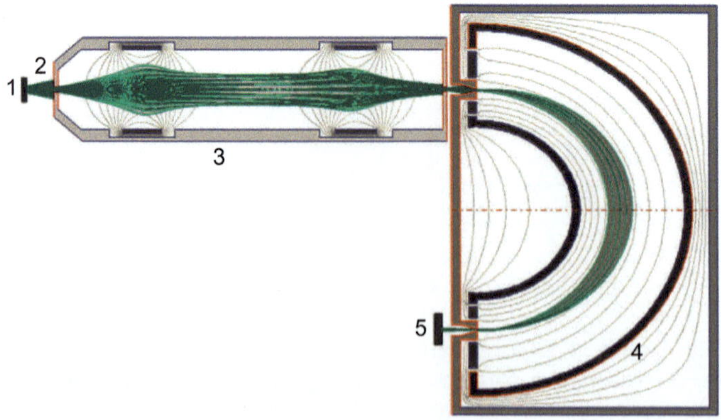

Chapter 1, Figure 5 A hemispherical deflector (4) with the input lens system (3): 1—sample, 2—secondary (photo) electrons, 5—collector (SEM).

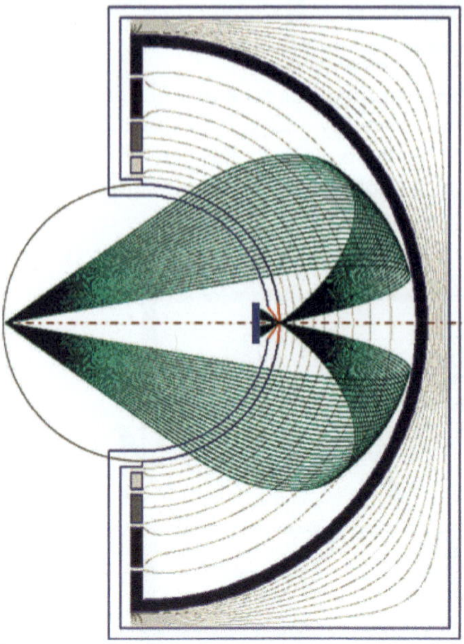

Chapter 1, Figure 6 Electron-optical scheme of a real spherical mirror with fringe-field correction.

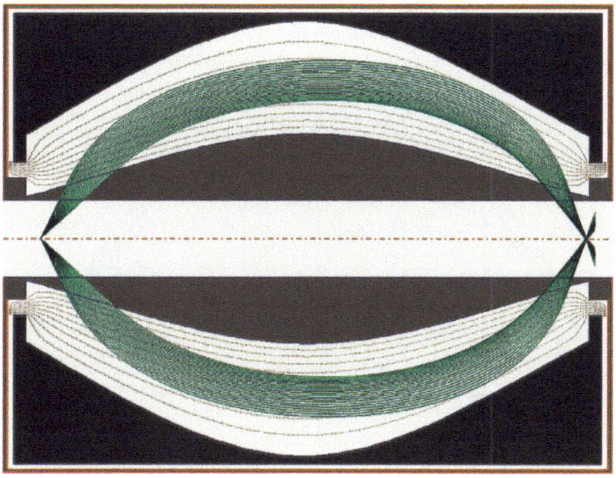

Chapter 1, Figure 8 Scheme of a real quasi-conical energy analyzer.

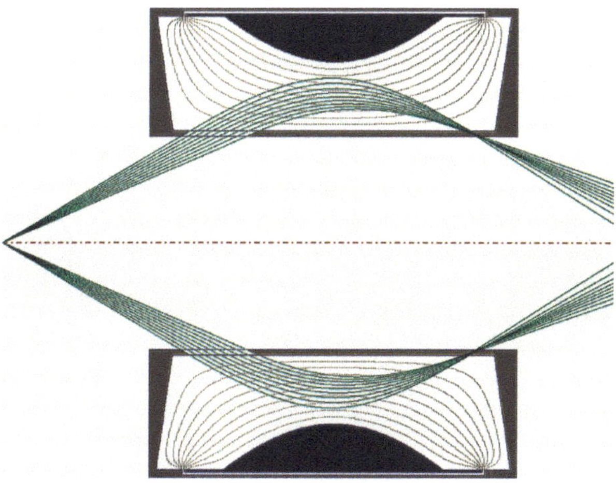

Chapter 1, Figure 9 Physically realizable approximation of an electron-optical scheme of the hexapole cylindrical analyzer with end-face electrodes at $\mu = 1$ and $\gamma = -1$ in the axis-ring mode of focusing.

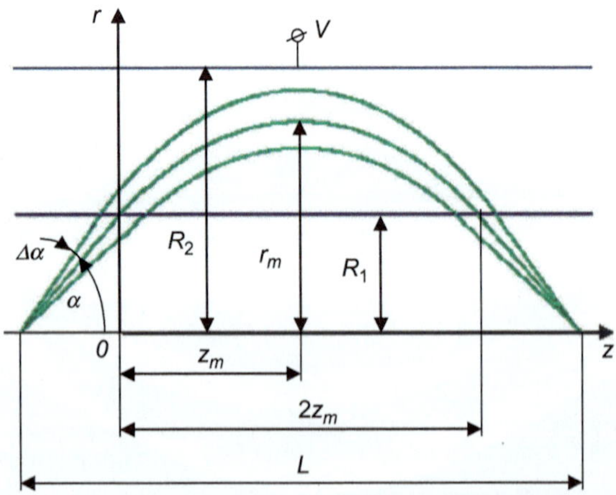

Chapter 2, Figure 1 Particle trajectories in a cylindrical capacitor.

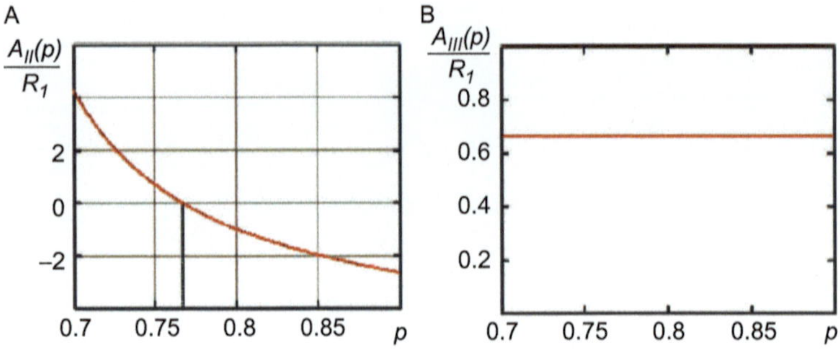

Chapter 2, Figure 2 The second (A) and third (B) coefficients of Eq. (11) as functions of the parameter p with the first-order focusing condition $A_I=0$ being fulfilled.

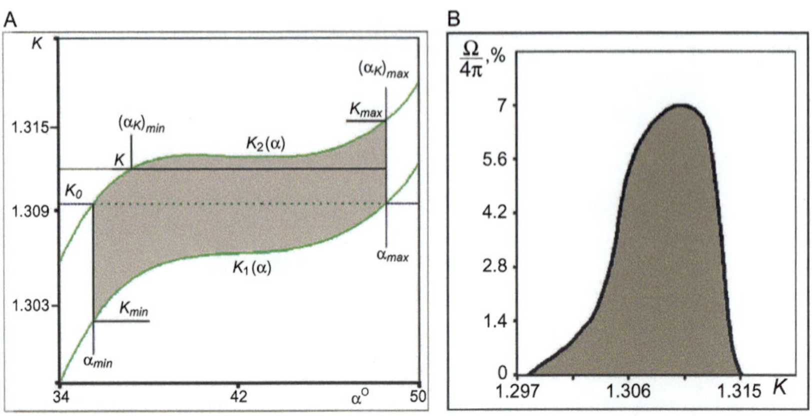

Chapter 2, Figure 4 Transmittance region (A) and instrumental function (B) of the energy analyzer in the mode of second-order focusing.

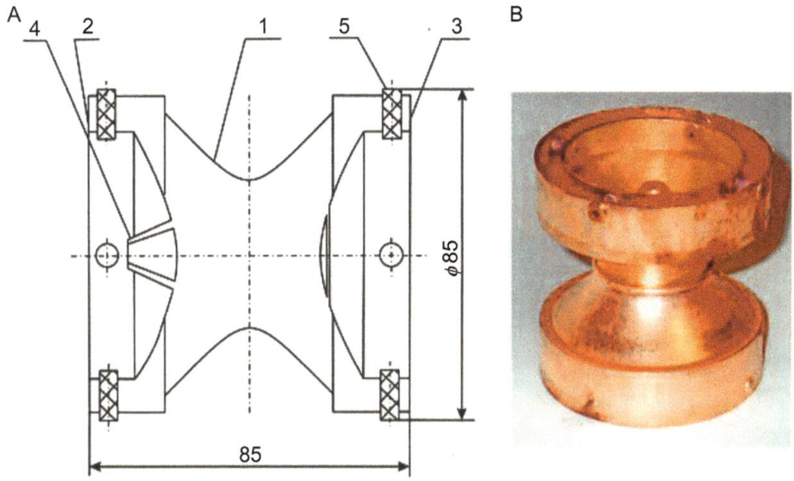

Chapter 2, Figure 54 Structure of electrodes (A) and external view (B) of the energy analyzer based on a complete hyperboloidal electrode system with $\gamma = 1.0$; $r_a = d_a = 19$ mm. 1, 2, 3 - field-generating electrodes, 4 – truncated cones, 5 – ceramic insulators.

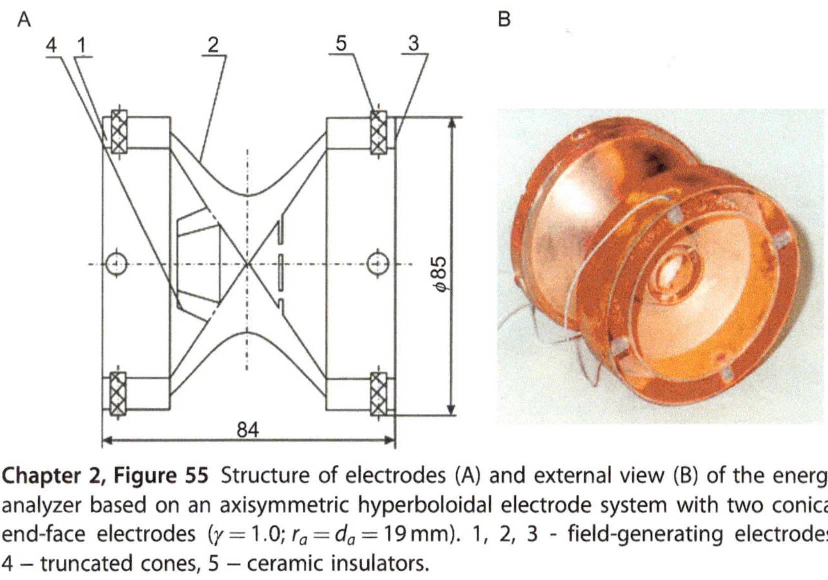

Chapter 2, Figure 55 Structure of electrodes (A) and external view (B) of the energy analyzer based on an axisymmetric hyperboloidal electrode system with two conical end-face electrodes ($\gamma = 1.0$; $r_a = d_a = 19$ mm). 1, 2, 3 - field-generating electrodes, 4 – truncated cones, 5 – ceramic insulators.

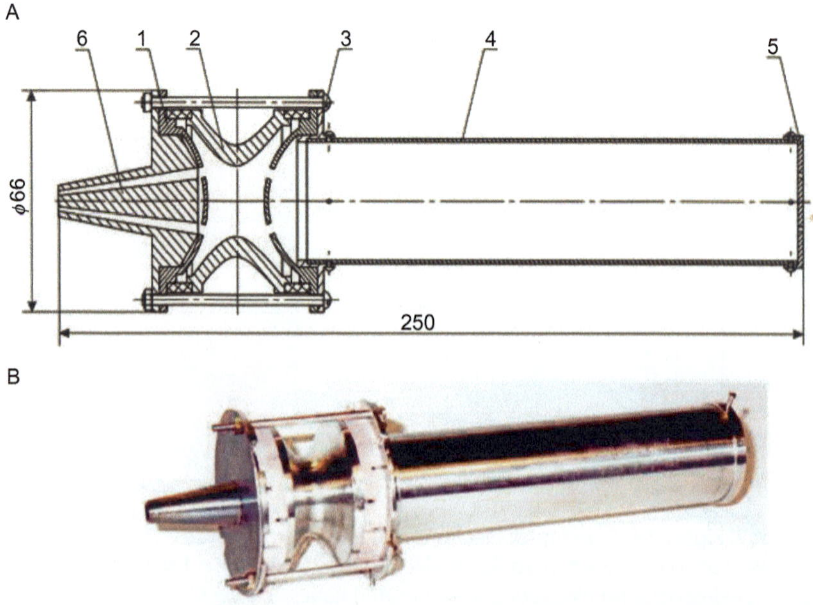

Chapter 2, Figure 56 Energy analyzer based on the hyperboloidal electrode system with selection slits located in the drift space ($\gamma=\sqrt{2}$; $r_a=11$ mm, $d_a=7.78$ mm). 1, 2, 3 – electrodes of hyperboloidal electrode system, 4 – drift tube, 5 – selection slits.

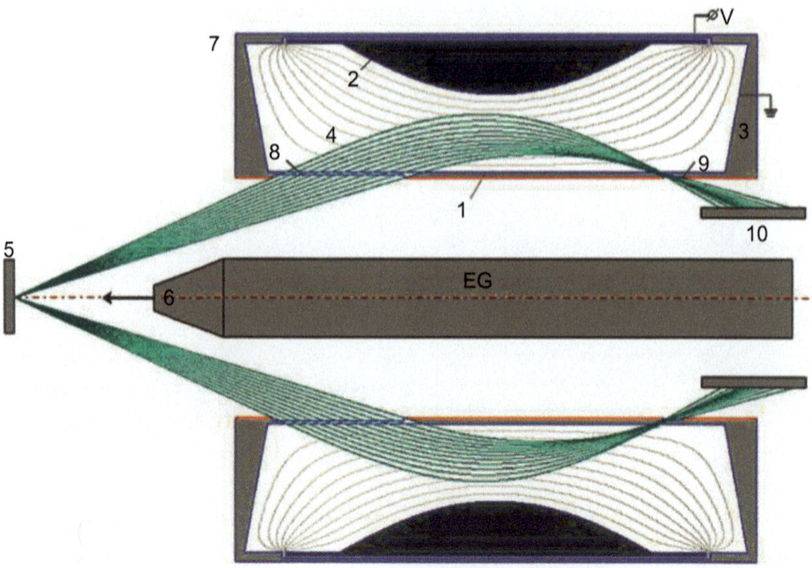

Chapter 3, Figure 4 Electron-optical scheme of the long-focus hexapole-cylindrical analyzer providing the mode of axis-ring angular focusing (1–cylindrical electrode, 2–outer electrode with a curved profile, 3–end-face electrodes, 4–secondary electrons, 5–sample, 6–primary electrons, 7–magnetic screen, 8 and 9–entrance and exit slits, 10–detector, EG–electron gun).

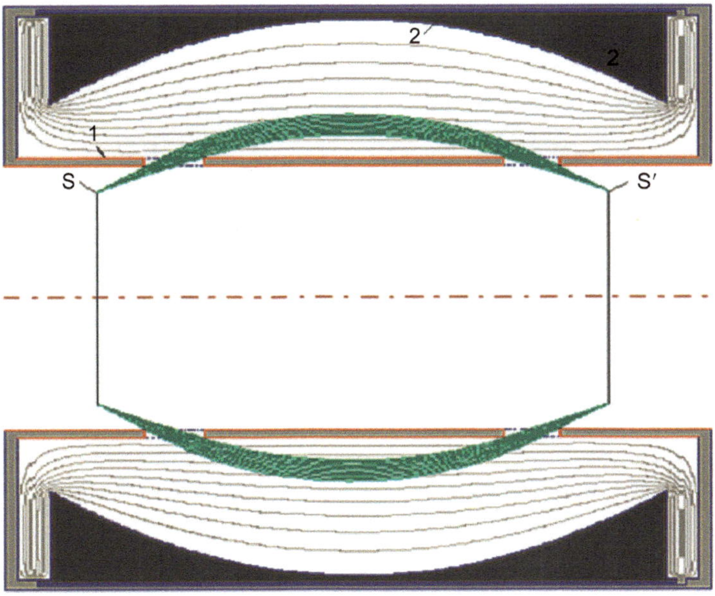

Chapter 3, Figure 6 Charged particle trajectories in the hexapole-cylindrical analyzer with $\gamma = 1$ (1—cylindrical electrode, 2—external deflecting electrode with a curved profile, *S*—annular source, *S′*—annular image).

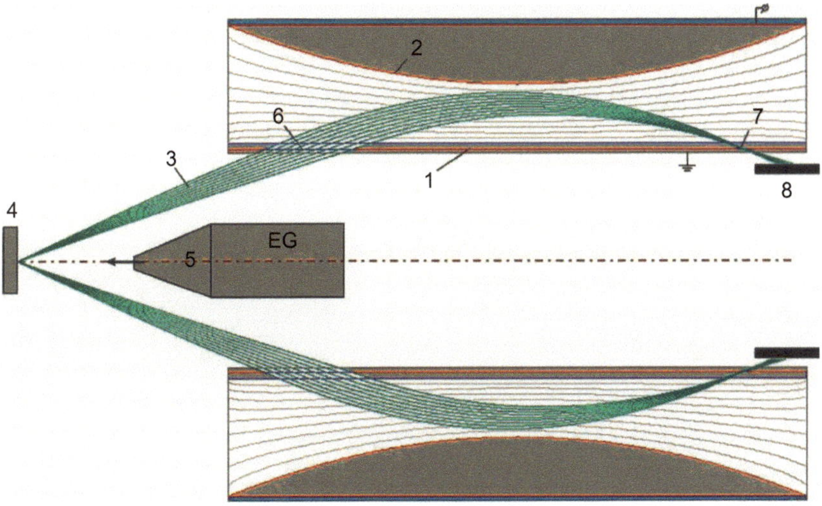

Chapter 3, Figure 8 Scheme of the hexapole-cylindrical energy analyzer in the mode of angular focusing axis-ring (1—cylindrical electrode, 2—external concave electrode, 3—secondary electrons, 4—test sample, 5—primary electrons, 6 and 7—input and output slits, 8—detector, EG—electron gun).

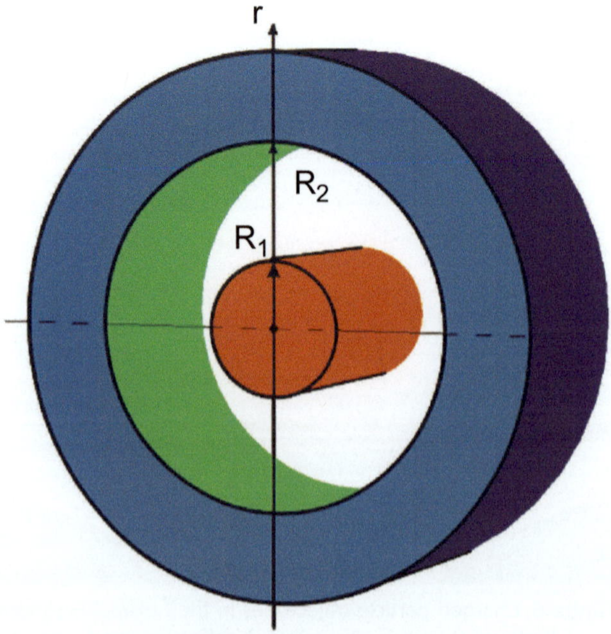

Chapter 4, Figure 6 Cylindrical capacitor with real-thickness electrodes.

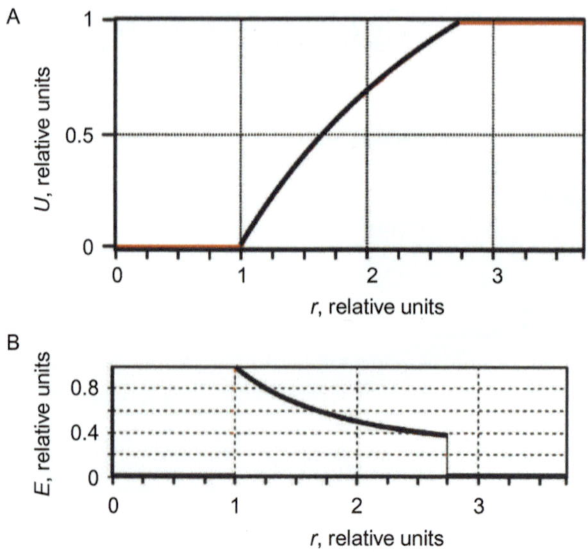

Chapter 4, Figure 7 Radial dependence of the potential distribution (A) and potential gradient's module (B) in the cylindrical capacitor with real-thickness electrodes.

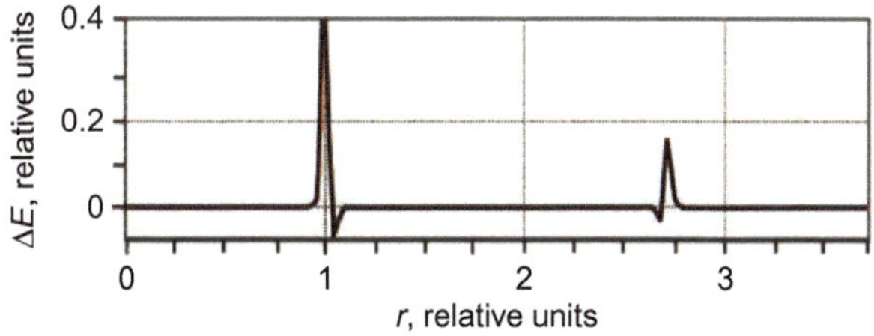

Chapter 4, Figure 8 The accuracy of differentiation using the finite-difference formulas.

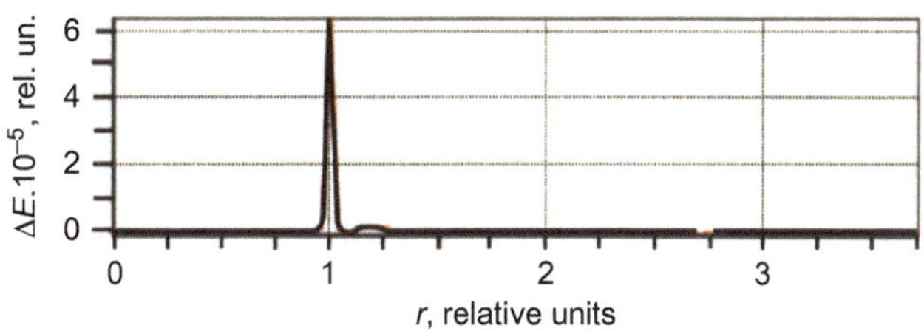

Chapter 4, Figure 11 Differentiation accuracy using the shifted finite-difference formulas.

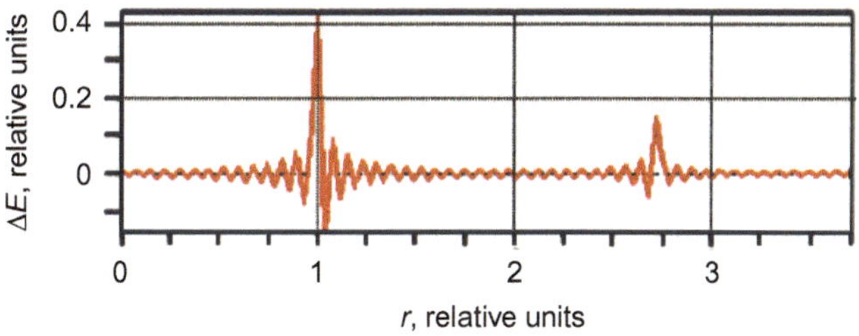

Chapter 4, Figure 12 Differentiation accuracy using the DFT.

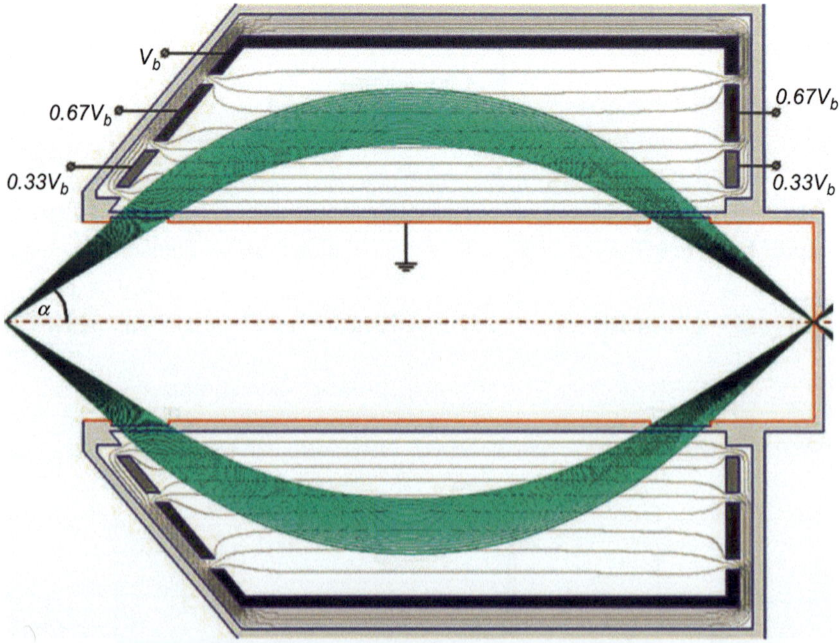

Chapter 4, Figure 15 Transaxial cross section of the energy analyzer, with two pairs of correcting rings, and electron trajectories with the energy E_0 $(E_0/V_b = 1.4311)$.

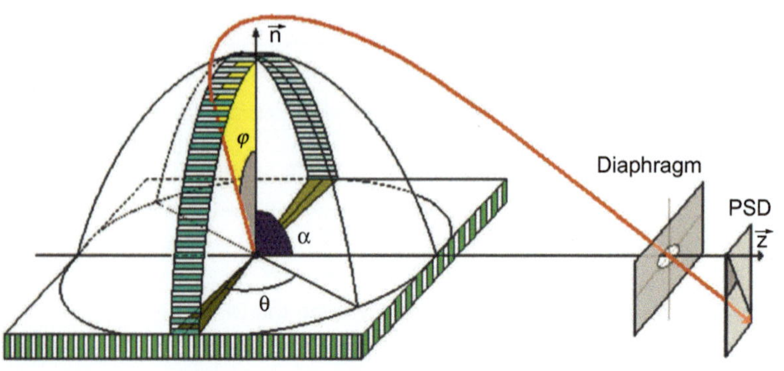

Chapter 4, Figure 17 Schematic of angular measurements.

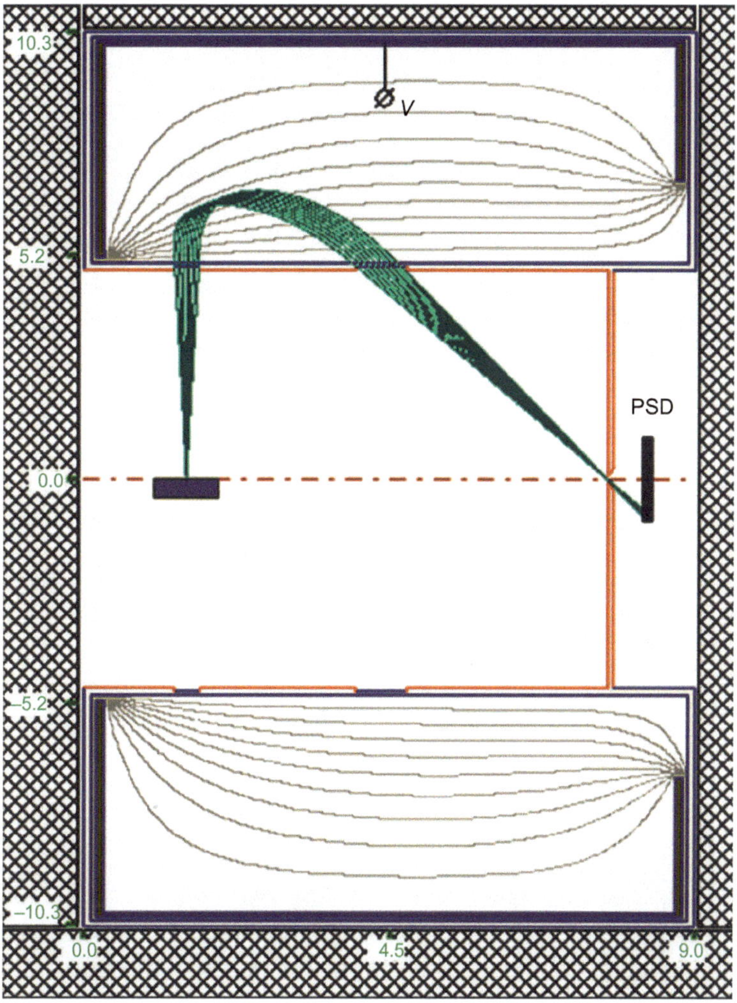

Chapter 4, Figure 18 Electron-optical scheme of the analyzer with energy and angular resolution: 1—input aperture, 2—output aperture, 3—position-sensitive detector with selectivity in the polar angle.

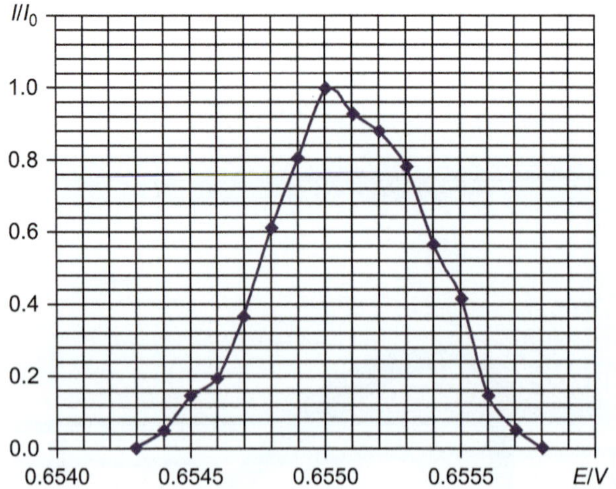

Chapter 4, Figure 19 Instrumental function of the analyzer with angular resolution.

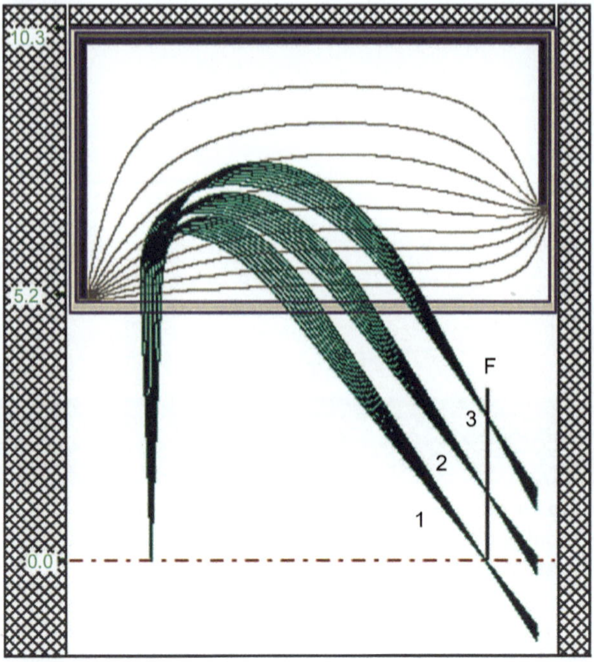

Chapter 4, Figure 20 Angular focusing of photoelectrons with different initial energies: 1—$E/V = 0.655$, 2—$E/V = 0.75$, 3—$E/V = 0.85$; *F*—focal plane.

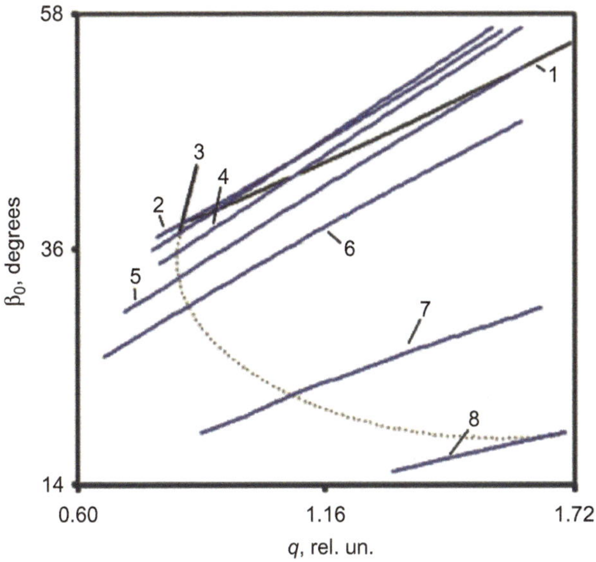

Chapter 4, Figure 22 Dependence of the second-order focusing angle on the normalized energy Angles θ_0: 1—10°, 2—20°, 3—30°, 4—40°, 5—50°, 6—60°, 7—90°, 8—110°.

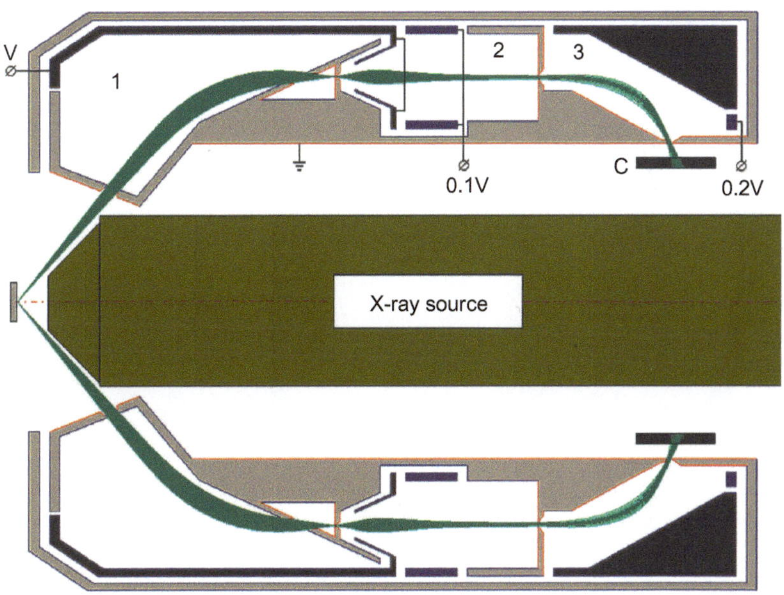

Chapter 4, Figure 25 Electron-optical scheme of a three-stage conical energy analyzer with an X-ray source.

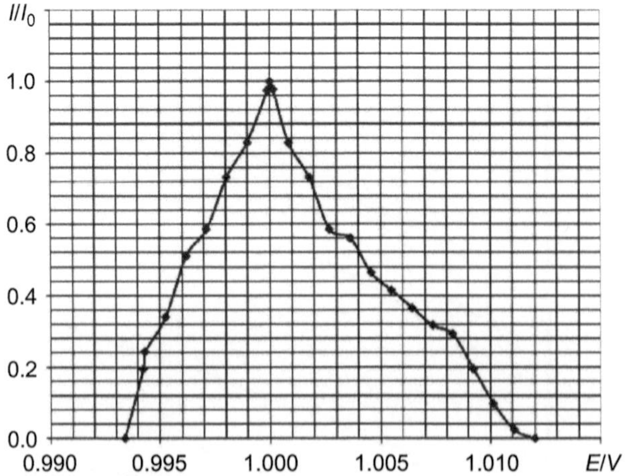

Chapter 4, Figure 26 Instrumental function of the conical analyzer.

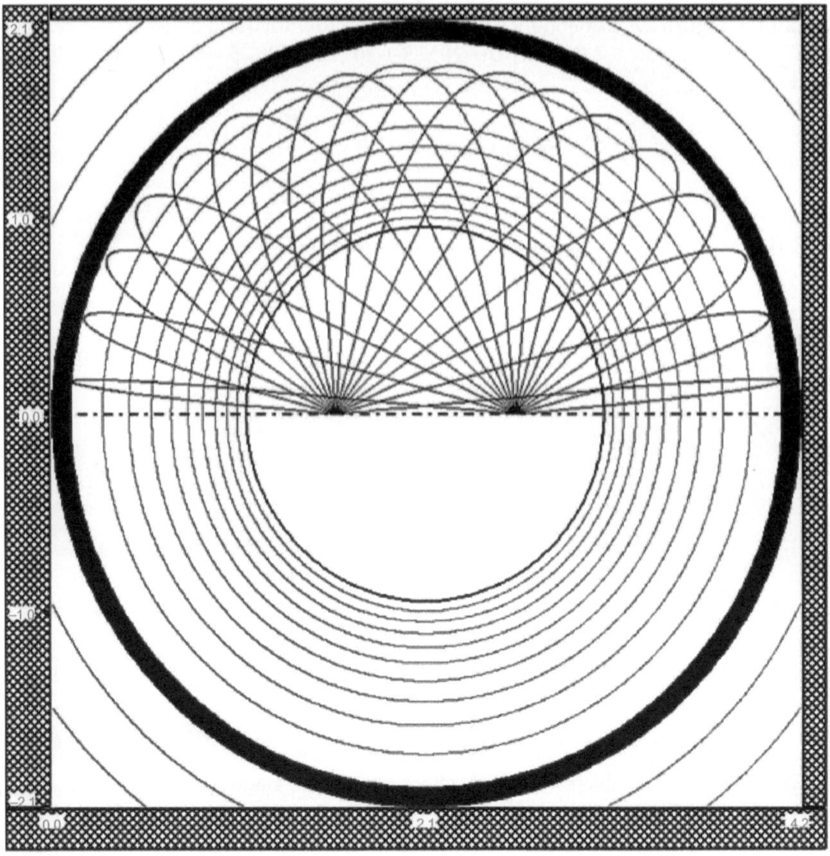

Chapter 4, Figure 27 Charged particle trajectories in a spherical mirror.

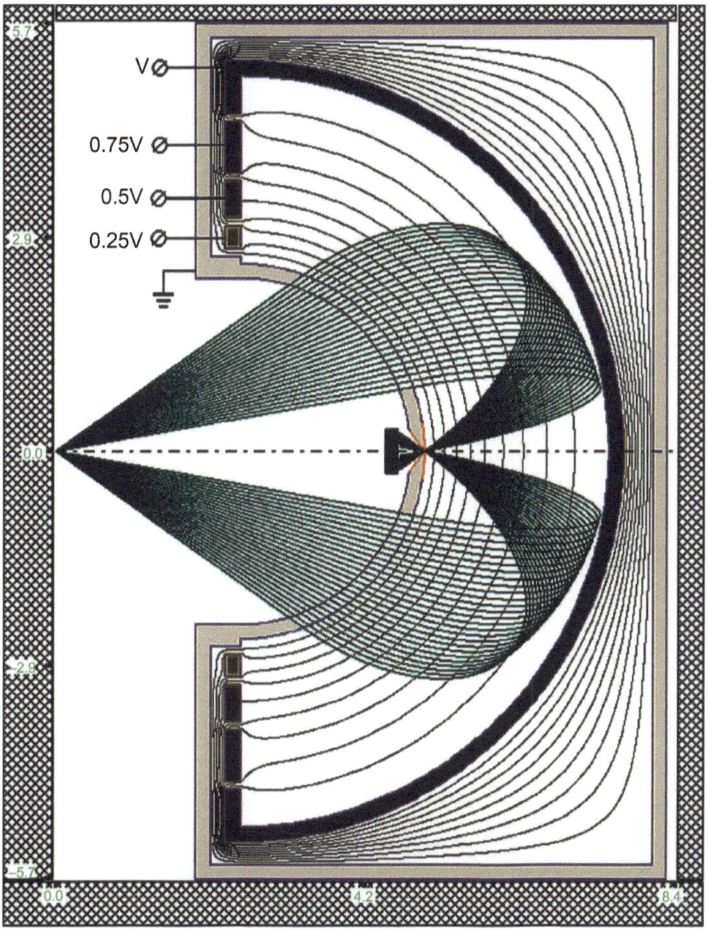

Chapter 4, Figure 28 Actual construction of the spherical mirror.

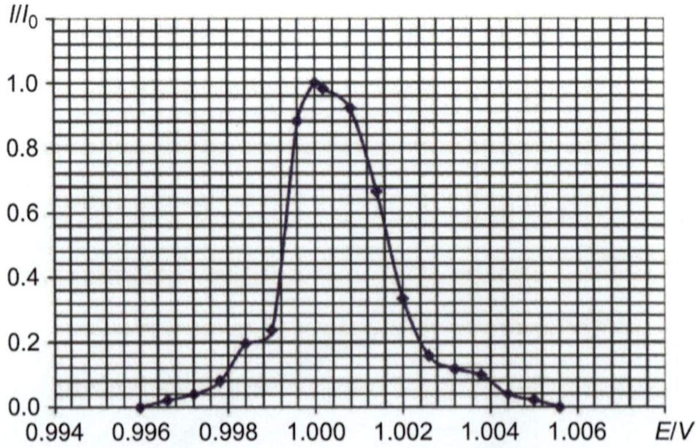

Chapter 4, Figure 29 Instrumental function of the spherical mirror.

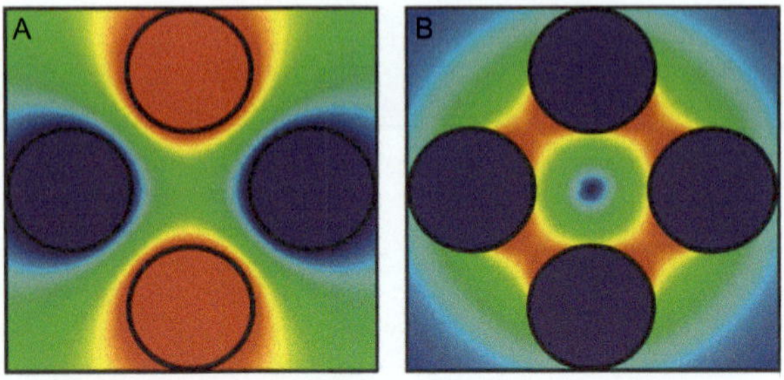

Appendices, Figure A3.4 Mapping of the electrostatic potential (A) and absolute value of the potential gradient (B) in a quadrupole system.

FSC
www.fsc.org
MIX
Papier | Fördert
gute Waldnutzung
FSC® C083411

Kolibri 360 GmbH
Ferdinand-Jühlke-Straße 7
99095 Erfurt, Deutschland
produktsicherheit@kolibri360.de